THE
BEHAVIORAL
AND
SOCIAL SCIENCES

Achievements and Opportunities

DEAN R. GERSTEIN, R. DUNCAN LUCE,
NEIL J. SMELSER, and SONJA SPERLICH, *Editors*

COMMITTEE ON BASIC RESEARCH IN
THE BEHAVIORAL AND SOCIAL SCIENCES
COMMISSION ON BEHAVIORAL AND SOCIAL SCIENCES AND EDUCATION
NATIONAL RESEARCH COUNCIL

with the cooperation of the

CENTER FOR ADVANCED STUDY IN THE BEHAVIORAL SCIENCES
and the
SOCIAL SCIENCE RESEARCH COUNCIL

NATIONAL ACADEMY PRESS
Washington, D.C.   1988

NATIONAL ACADEMY PRESS   2101 Constitution Avenue, NW   Washington, DC 20418

NOTICE: The project that is the subject of this report was approved by the Governing Board of the National Research Council, whose members are drawn from the councils of the National Academy of Sciences, the National Academy of Engineering, and the Institute of Medicine. The members of the committee responsible for the report were chosen for their special competences and with regard for appropriate balance.

This report has been reviewed by a group other than the authors according to procedures approved by a Report Review Committee consisting of members of the National Academy of Sciences, the National Academy of Engineering, and the Institute of Medicine.

The National Academy of Sciences is a private, nonprofit, self-perpetuating society of distinguished scholars engaged in scientific and engineering research, dedicated to the furtherance of science and technology and to their use for the general welfare. Upon the authority of the charter granted to it by the Congress in 1863, the Academy has a mandate that requires it to advise the federal government on scientific and technical matters. Dr. Frank Press is president of the National Academy of Sciences.

The National Academy of Engineering was established in 1964, under the charter of the National Academy of Sciences, as a parallel organization of outstanding engineers. It is autonomous in its administration and in the selection of its members, sharing with the National Academy of Sciences the responsibility for advising the federal government. The National Academy of Engineering also sponsors engineering programs aimed at meeting national needs, encourages education and research, and recognizes the superior achievements of engineers. Dr. Robert M. White is president of the National Academy of Engineering.

The Institute of Medicine was established in 1970 by the National Academy of Sciences to secure the services of eminent members of appropriate professions in the examination of policy matters pertaining to the health of the public. The Institute acts under the responsibility given to the National Academy of Sciences by its congressional charter to be an adviser to the federal government and, upon its own initiative, to identify issues of medical care, research, and education. Dr. Samuel O. Thier is president of the Institute of Medicine.

The National Research Council was organized by the National Academy of Sciences in 1916 to associate the broad community of science and technology with the Academy's purposes of furthering knowledge and advising the federal government. Functioning in accordance with general policies determined by the Academy, the Council has become the principal operating agency of both the National Academy of Sciences and the National Academy of Engineering in providing services to the government, the public, and the scientific and engineering communities. The Council is administered jointly by both Academies and the Institute of Medicine. Dr. Frank Press and Dr. Robert M. White are chairman and vice chairman, respectively, of the National Research Council.

Photograph credits: (pages 6 and 238) © Kenneth Garrett/Woodfin Camp; (page 48 and 238) © Rick Brady 1985/Uniphoto; (page 52) courtesy of Dr. Paul Ekman; (pages 84 and 238) © Bob Daemmrich/Uniphoto; (pages 128 and 238) Tildon Easton Pottery Kiln, courtesy of Alexandria Archaeology, City of Alexandria, Virginia; (pages 166 and 238) © Kenneth Garrett 1982/Woodfin Camp; (pages 202 and 238) courtesy of Reed College.

Library of Congress Catalog Card Number 88-1618

ISBN 0-309-03749-2

Copyright © 1988 by the National Academy of Sciences

Printed in the United States of America

Committee on Basic Research in the Behavioral and Social Sciences

R. DUNCAN LUCE *(Cochair)*, Department of Psychology and Social Relations, Harvard University

NEIL J. SMELSER *(Cochair)*, Department of Sociology, University of California, Berkeley

MEINOLF DIERKES, Science Center Berlin, Federal Republic of Germany

JOHN A. FEREJOHN, Department of Political Science, Stanford University

LAWRENCE M. FRIEDMAN, School of Law, Stanford University

VICTORIA FROMKIN, Graduate Division and Department of Linguistics, University of California, Los Angeles

ROCHEL GELMAN, Department of Psychology, University of Pennsylvania

LEO A. GOODMAN, Department of Statistics and Department of Sociology, University of Chicago, and University of California, Berkeley

JAMES G. GREENO, School of Education, Stanford University

EUGENE A. HAMMEL, Graduate Group in Demography and Department of Anthropology, University of California, Berkeley

LEONID HURWICZ, Department of Economics, University of Minnesota

EDWARD E. JONES, Department of Psychology, Princeton University

GARDNER LINDZEY, Center for Advanced Study in the Behavioral Sciences, Stanford, California

DANIEL L. McFADDEN, Department of Economics, Massachusetts Institute of Technology

In Memoriam

———

Glynn Llewelyn Isaac
1937—1985

The committee was greatly aided in its work by groups of colleagues who prepared background papers for this volume. Glynn Isaac, professor of anthropology at Harvard University, chaired one such group. After transmitting his group's manuscript, he became seriously ill while working overseas, and he died in Yokosuka, Japan, en route home. We honor here his contributions to this project and to the science of human origins.

Contents

4

INSTITUTIONS AND CULTURES / 129

5

METHODS OF DATA COLLECTION, REPRESENTATION, AND ANALYSIS / 167

Preface

This is a report on scientific frontiers in the behavioral and social sciences—leading research questions and fundamental problems—and on the new resources needed to work on them.

This volume is a successor to two earlier studies by the Committee on Basic Research in the Behavioral and Social Sciences. In one, *Behavioral and Social Science: Fifty Years of Discovery* (1986), we scanned the work of the past, identifying specific lines of accumulated knowledge and broad shifts in emphasis since the 1933 report of the President's Research Committee on Social Trends. In the other, *Behavioral and Social Science Research: A National Resource* (1982), we considered particular cases and presented our judgments concerning the present value, significance, and social utility of basic research in these disciplines.

Against this backdrop, the current volume looks to the future. When this phase of the committee's work was first envisioned early in 1983, there was a clear federal policy of steadily rising science budgets tailored to specific research initiatives. Accordingly, we were asked by the National Science Foundation, the committee's initial sponsor, to help define some discrete priorities for increased investments in behavioral and social sciences research, which would be comparable to the priorities recommended by groups representing other fields of science, such as the National Research Council (NRC) "outlook" report, *Astronomy and Astrophysics for the 1980's* (1982). However, that report and several more recent NRC reports of the same genre, including *Renewing U.S. Mathematics* (1984), *Opportunities in Chemistry* (1985), and *Physics Through the*

1990s (1986), have dealt with a single scientific discipline. We were asked to represent all the behavioral and social sciences, a highly diverse congregation of separate disciplines. The task was not an easy one, and we can imagine that a different group of researchers might have taken a different approach to it than the largely interdisciplinary one that we chose.

The sponsorship of our study has broadened to include seven additional public and private agencies with differing missions and interests, reflecting the diversity of concerns and sources of support for behavioral and social sciences research: National Institute on Aging, National Institute of Child Health and Human Development, National Institute of Mental Health, U.S. Army Research Institute for the Behavioral and Social Sciences, Russell Sage Foundation, System Development Foundation, and the National Research Council Fund.* As a result of our multidisciplinary scope and breadth of sponsorship, the initial charge of defining priorities for the investment of incremental funds was extended to include consideration of the general institutional conditions and support system for behavioral and social sciences research.

From the outset the committee members recognized that we could not carry out the task by ourselves. A very important part in enlarging participation was played by two organizations that have formally cooperated with the NRC in the study: the Center for Advanced Study in the Behavioral Sciences and the Social Science Research Council. With their assistance, we identified some 2,400 scientists, including both established and young behavioral and social sciences researchers, and asked them about their part of the research enterprise: Where is it heading with respect to intellectual ferment, the generation of empirical discoveries, and major theoretical and methodological developments? We also asked them to identify key researchers to help the committee examine these areas of ferment. We further broadcast our appeal for assistance to 150 journals.

We received detailed replies from about 600 researchers, who identified more than 1,000 topics or lines of research, many of them overlapping, and gave us more than 2,000 names to consider. The committee worked carefully and critically through this mass of advice, rejecting some ideas that appeared idiosyncratic or marginal and seeking common threads among the others. While some suggestions found rather little reflection in the ultimate course of

---

*The National Research Council Fund is a pool of private, discretionary, nonfederal funds that is used to support a program of Academy-initiated studies of national issues in which science and technology figure significantly. The NRC Fund consists of contributions from a consortium of private foundations including the Carnegie Corporation of New York, the Charles E. Culpeper Foundation, the William and Flora Hewlett Foundation, the John D. and Catherine T. MacArthur Foundation, the Andrew W. Mellon Foundation, the Rockefeller Foundation, and the Alfred P. Sloan Foundation; the Academy Industry Program, which seeks annual contributions from companies that are concerned with the health of U.S. science and technology and with public policy issues with technological content; and the National Academy of Sciences and the National Academy of Engineering endowments.

the study and others were highly influential, the committee is indebted to and appreciative of everyone who responded to our call for assistance.

Ultimately, we selected 31 topics as the basis for working groups. Early in 1985 we gave the groups (which had from 5 to 11 members) all of the information and advice we had garnered with respect to the relevant topical areas. Some 6 months later we received back 31 concise papers on research opportunities and needs. These working papers very much informed and influenced this report. We take pleasure in acknowledging the generous assistance that the members of the working groups—especially their chairs—gave us in this study, and we record their names, with our thanks, in Appendix B. For readers interested in exploring more intensively the topics discussed in this report, the Russell Sage Foundation is currently preparing for publication a volume of those papers, which includes specific references to the large underlying scientific literature.

One major issue the committee faced was whether to organize this report along conventional disciplinary lines—to prepare separate chapters about anthropology, economics, political science, psychology, sociology, and so forth— or to adopt some other scheme of organization. Disciplines are, to be sure, the basis on which academic departments in universities and colleges are usually organized, the structure under which the bulk of fundamental behavioral and social sciences research, training, and instruction is conducted, and the arena in which most scientific careers are made. Major professional associations are also organized by traditional disciplines, as is a large fraction of funding by research agencies. There is also ample precedent for a disciplinary approach, most prominently the "BASS" report, *The Behavioral and Social Sciences: Outlook and Needs* (1969) and its companion volumes, prepared by the predecessor committee most comparable to ours.

Notwithstanding these precedents and conventions, our committee from the beginning favored another approach. We did so partly from a sense that many of the opportunities currently visible in the behavioral and social sciences spring from and support the development of methods, tools, and concepts across disciplines. The topics and lines of research mentioned in our initial survey confirmed very strong interdisciplinary themes, and when we formed working groups, the great majority were interdisciplinary in composition. Finally, the recommendations regarding resource needs that emerged from the working papers and our further deliberations were far more inclined to cross the boundaries between disciplines than to be delineated by them. The interdisciplinary note is strong in all that follows.

All National Research Council reports are subject to review by an expert group other than the authors. In this instance, the review process has been more extensive than most. The boards of the Center for Advanced Study in the Behavioral Sciences and the Social Science Research Council participated fully with the Commission on Behavioral and Social Sciences and Education in the review process. In addition, the draft manuscript was reviewed by the chairs of the 31 working groups, by scientists selected by the NRC Report Review Committee, and by others at the request of our committee. We are grateful to the many colleagues who read and formally commented on this report; their insightful critiques enabled us to improve it substantially. We have striven to use their advice, along with that of the many other colleagues who have written and spoken informally to us in the course of the enterprise, to more faithfully represent the full range of knowledge and perspectives bearing on our task.

We are indebted to all of the public and private agencies sponsoring this project for their encouragement, cooperation, and support. Among the many officials who have been important in our efforts, we wish especially to acknowledge the energy and vision of the former senior associate for behavioral and social sciences at the National Science Foundation, Otto N. Larsen, and the assistant director for biological, behavioral, and social sciences, David A. Kingsbury.

The former executive director of the Commission on Behavioral and Social Sciences and Education, David A. Goslin, provided experienced advice, analytic intelligence, and administrative backing. The Commission's associate director for reports, Eugenia Grohman, read successive drafts, joined committee discussions, and gave us many useful suggestions for revising the report and polishing the text. The behavioral and social sciences are fortunate to draw on the unusual talents and exacting standards of these two individuals.

We would also like to acknowledge those researchers whose guidance, published work, or other assistance enabled us to develop illustrations: Martin Baily, Patricia Carpenter, Paul Ekman, Robert Hall, Reid Hastie, Marcel Just, William Labor, Ian Madieson, James McClelland, Charles Nelson, and Herbert Pick.

Every committee member participated in the original drafting of the report, but we would like to express particular appreciation to John Ferejohn, Rochel Gelman, Leo Goodman, Eugene Hammel, and Barbara Rosenkrantz, who chaired drafting subcommittees. The tasks of organizing and shaping these texts and revising and completing the report were undertaken by the cochairs and the committee's professional staff, study director Dean R. Gerstein and senior research associate Sonja Sperlich. These two staff members, along with the committee's administrative secretary, Linda B. Kearney, her predecessor, Beverly R. Blakey, and assistant William A. Vaughan, Jr., also managed the administrative and logistical requirements of the study, an organizational effort spanning 3 years, scores of meetings, hundreds of participants in committee, working group and review activities, and forests of correspondence.

The national community of behavioral and social scientists is far too large and diverse—with more than 100,000 PhD's in more than a dozen disciplines—for there to be complete concordance in a single document. This is especially the case with regard to selecting for explicit mention a limited number of promising research opportunities. The issue here is not so much controversy over any particular selection as the realization that others might also have been singled out. The research opportunities discussed here are a purposive sample from a larger universe of such opportunities. But to the degree that it is a good sample, the resulting recommendations for strengthening the research support system and raising the scientific yield can be considered to speak for and serve the best interests of that larger universe.

R. Duncan Luce and Neil J. Smelser, *Cochairs*
Committee on Basic Research in the
Behavioral and Social Sciences

THE
BEHAVIORAL
AND
SOCIAL SCIENCES

Introduction

The behavioral and social sciences strive to understand the conduct of human beings and animals, singly and in groups, from the moments of their birth to the moments of their death. The subject matter of these sciences ranges from global commerce and conflict to the neurochemical substrates of memory and motivation. The interests of research carry from the origins of species to forecasts of political, economic, and technological behavior and events. Psychological, social, and cultural studies pertain to virtually everything that people treat as a problem in our civilization—violence, theft, pollution, and illness—and nearly everything hailed as a triumph—justice, plenitude, artistry, and freedom. Even in events that are nominally quite technical in character, such as the eradication of polio or the explosion of the space shuttle, human factors, behavioral and organizational, play a large role.

The proximate goal of the behavioral and social sciences is to discover, describe, and explain behavioral and social phenomena in accord with the canons of scientific logic and method. As in all sciences, curiosity and imagination, the desire to understand events and the capacity to create ideas, strongly motivate research on behavior and society. Motivation also comes from demands for practical knowledge that can shape and improve behavior and advance social purposes, demands fueled by past achievements and encouraged by promises of future returns. Modern life has been and continues to be profoundly transformed—more than is commonly realized—by such widely diffused innovations as standardized tests, probability sampling methods, longitudinal survey techniques, and industrial quality-control methods. It has also been profoundly changed by an understanding of visual and auditory perception, the aging process, the roles of families and schools in the development of cognitive skills, the causes of gender and racial differences, the dynamics of inflation, and the nature of poverty and dependency.

The harvest of the behavioral and social sciences has been rich at the same time that the modes of cultivation have undergone far-reaching shifts. Before the 1960s, research in these sciences was largely a project of universities and private foundations. A major but temporary addition to this partnership was the national mobilization of behavioral and social researchers from 1942 to 1945 to help manage the wartime economy, train and assign millions of soldiers and sailors, analyze and outwit the enemy, and, finally, devise plans for peacetime demobilization and reconstruction. This highly successful effort was not extended in the immediate post-war period. With some exceptions (for example, the Council of Economic Advisers, RAND, and certain programs of the Office of Naval Research), the professors returned to their chairs, and federal research support focused largely on the physical and life sciences.

In the late 1950s and early 1960s, a major change occurred: the federal government significantly expanded and private foundations dramatically reduced support for basic and applied research, including advanced training, in the behavioral and social sciences. The federal government became a direct producer of such research as well as the dominant patron for research in universities and colleges, free-standing research institutes, and a variety of contractors specializing in program evaluation.

In the past 15 years, the federal role has diminished in many respects and no other sector has compensated. This change is perhaps best represented in the shrinkage of federal support. Between fiscal 1972 and fiscal 1987, federal support for behavioral and social sciences research decreased by about 25 percent in constant dollars; during the same period, constant-dollar federal support for other fields of scientific research increased by about 36 percent. Of the total federal budget for research, the percentage for behavioral and social sciences declined from 8.0 percent in 1972 to 4.6 percent in 1987; in constant (1987) dollars, from $1.03 billion to $0.78 billion.

The United States has been in the forefront of virtually every behavioral and social sciences field since the 1960s. But if the funding trends of recent years persist, many of these leading roles are likely to be taken by other countries in the 1990s. Cross-national statistics suggest that Japan and, to a lesser extent, some European countries now spend substantially larger proportions of their national resources on behavioral and social sciences research than does the United States. In the light of past contributions, but even more so in the light of present opportunities for scientific advances with their concomitant yield of practical benefits, it is time for a new national commitment to the behavioral and social sciences in the United States.

Although the federal government is the major source of support for behavioral and social sciences research, that support is quite decentralized. It comes from many different departments and agencies, which have divergent interests and missions as well as different methods of funding: for example, grants from research agencies, such as the National Institute of Mental Health, and contracts from operational agencies, such as the U.S. Department of Labor. This decentralization has some benefits: for example, it corresponds to the great diversity of research interests across the behavioral and social sciences. But it also accounts in part for the lack of consideration of research support as a whole, marked fluctuations in levels of support, and resistance to planning or sustaining a general strategy to keep the behavioral and social sciences strong.

Just as research support in the federal government is divided among many, diverse agencies, researchers themselves inhabit a decentralized realm. The behavioral and social sciences consist of distinct disciplines that typically constitute separate departments in colleges and universities as well as separate associations of professionals and scientists. The original core of disciplines, which emerged in the late nineteenth century, included anthropology, economics, political science, psychology, and sociology. Their number has since grown. Linguistics and geography have been added to the original core. History is now to a large degree a social science, from the standpoint of both subject matter and methods of investigation and explanation. The same can be said of much research in business, education, law, and psychiatry. More specialized, newer fields, such as artificial intelligence, child development, cognitive science, communications, demography, and management and decision science, involve one or more of the classic behavioral and social sciences. Mathematics, computer science, and, especially, statistics find many applications and connections to research conducted by behavioral and social scientists.* In summary, the disciplinary foundations of the behavioral and social sciences are strong. Academic departments, professional associations, and journals have attended capably to instruction, professional accreditation, and identification of professional opportunities and needs.

The disciplines as such are not the main concern of this report. Instead, we have tried to track and spotlight some of the most important and promising lines of research in the behavioral and social sciences as a whole, to project a vision of how such lines might develop in the near future, and to specify the resources and organizational arrangements that can contribute to achieving their greatest scientific potential over that term.

Chapters 1 through 5 of this report characterize the substantive research

---

*For a more detailed account of the disciplinary structure of the behavioral and social sciences, see Chapter 2 of *Behavioral and Social Science Research: A National Resource, Part I*, Robert McC. Adams, Neil J. Smelser, and Donald J. Treiman, eds. Washington, D.C.: National Academy Press, (1982).

opportunities that lead most directly to our recommendations. Chapter 1, "Behavior, Mind, and Brain," focuses on research concerning individual behavioral and mental processes of sensory perception, memory and learning, cognition, and language. Chapter 2, "Motivational and Social Contexts of Behavior," considers affective states and processes, the linkages between health, behavior, and social context, the causes and control of violent crime, and the nature of social interaction. Chapter 3, "Choice and Allocation," deals with research on individual and collective decisions and their consequences in market-based and other economic systems, contracts, organizational hierarchies, and occupational systems. Chapter 4, "Institutions and Cultures," considers more global and historical aspects of society—including human evolution, demography, modernization processes, science and technology, world trade, and international conflict. Chapter 5, "Methods of Data Collection, Representation, and Analysis," concerns methodological research, which sharpens the observational and explanatory powers of the behavioral and social sciences.

The committee considers this selection of substantive topics to be a very good sample of high-quality and promising work in the behavioral and social sciences. The particular results and studies noted are illustrative, not exhaustive. We have necessarily not discussed numerous other lines of research that may turn out to be comparable in excellence and importance to those included. Moreover, the new and highly original are by their very nature not easily anticipated. We certainly do not advocate that only the lines of research specifically cited here should be supported and all others given low priority. Detailed decisions about which specific research projects, fellowships, institutes, equipment, data collections, or centers to support—and especially decisions about which not to support—will best emerge from the continuing review by qualified scientists of well-executed studies, proposals, and applications. Priority-scored review by disinterested committees of researchers is a form of competitive selection that cuts off less-compelling research investments with, if anything, an excess of ruthlessness.

The last two chapters of this report step back from the substantive specifics to give an overall perspective on the scientific enterprise and its conditions. Chapter 6, "The Research Support System," deals with the human, technological, data-generating, and funding resources that shape opportunities in the behavioral and social sciences. This chapter draws together our conclusions on how to make the support system stronger and more productive. Chapter 7, "Raising the Scientific Yield," summarizes the research opportunities and new initiatives that we recommend. (Appendix A supplements this chapter with analyses of trends in the scale of federal and private foundation research funding.) We believe that this agenda of activities will go far toward ensuring the vitality and preeminence of the national effort in behavioral and social sciences research.

1

Behavior, Mind, and Brain

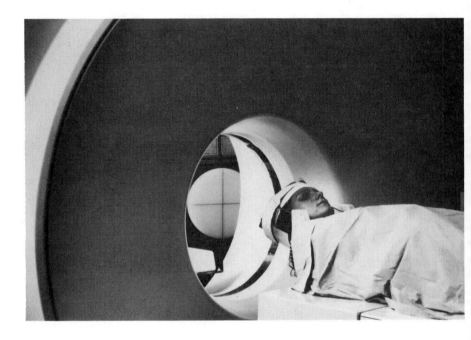

1

Behavior, Mind, and Brain

From the beginnings of scientific inquiry, researchers have tried to understand the workings of the mind and its relationship to behavior. In modern terms, scientists seek to answer such questions as:

How does an individual manage to see coherent objects in changing patches of multicolored light or to hear speech and music in bursts of sound varying in loudness and pitch?

How do people remember, even imperfectly, the vast storehouse of factual and functional information that each of us carries about?

How do infants—and other nonverbal creatures—think, and what are their thoughts like?

How does an individual learn new ideas, create concepts, organize his or her knowledge, and act upon what he or she knows?

Until about 100 years ago, these and similar questions led mainly to speculation, to sophisticated but untestable guesses about the nature of the mind and its function in behavior. Many scientists thought that this confinement to speculation would always be the case, on the presumption that mental life, unlike the world of objects, is not directly observable and so is not amenable to systematic, scientific study. But during the second half of the nineteenth century, experimental and close observational approaches to such questions began to appear. Some of the first work involved the study of sensory processes that underlie organic sensitivity to different physical stimuli, rigorous observations about how behavior is shaped by experience, and the relation of observable behaviors to certain brain regions and neural events.

Both theory and method have since advanced at an accelerating pace, with extensive revisions of earlier questions to make them more tractable and to take advantage of the growing linkages between knowledge about behavior, the mind, and the brain. Instead of asking only how best to measure the intelligence of people, researchers now ask about the nature of intelligent action, whether exhibited by humans, animals, computers, or robots. In addition to seeking laws that directly relate external stimuli to behavioral responses, researchers now attempt to unravel the basic processing that the brain must carry out in order to generate behavior and to simulate that processing on computers to formalize and test theories. These are the new questions—with their extensions into the details of visual and auditory processing, memory formation and retrieval, language, cognition, and action—that drive the research investigations highlighted in this chapter.

SEEING AND HEARING

As you look around, you see a variety of objects at various distances, some still and others moving, some transparent and some densely colored, some partially obscured and some not. All is perfectly ordinary and seen without effort if the light is adequate. Your cat or dog sees these things, too, although somewhat differently from the way you do. And so does the fly that evades your swat. And when you listen to a musical recording, you have little trouble hearing the separate instruments. If someone speaks while you are listening to the music, you have no difficulty in understanding the words, unless the music is very loud or you suffer from a certain form of hearing loss that is common in older males.

It is all so commonplace—but no one yet knows exactly how the brain does any of these things. No one yet knows enough to program a computer to pick out a wide range of objects from a scene, to isolate a violin in an orchestral passage, or to separate speech from noise and to partition it into words. Enabling machines to do these things would surely affect the way people live as much as the telephone, the thermostat, or the radar have done. And it is clear these tasks can be done because the human brain does them, continuously and apparently effortlessly. One approach to studying how these tasks are carried out is to work with computers and sensing devices, without much regard for how the tasks are done in the brain. Another approach is to focus directly on unraveling nature's way of doing them.

These tasks may be very complex. Or they may be simple but involve principles that scientists do not yet understand. For example, between one-quarter and two-fifths of the total cortex of the human brain is devoted to vision, suggesting that it is one of the most complex brain functions. However, abilities somewhat similar to human vision are exhibited by animal brains far more modest. For example, pigeons trained to identify the occurrence of trees (in contrast to bushes or other plants) in color slides—whether alone or in a forest, in whole or in part, leafy or bare—are then able to identify trees as accurately as people can, in slides they have not previously seen. Pigeons can also be trained to recognize a particular person in slides of individuals or crowds. Yet, with similar training, pigeons cannot identify simple line drawings, such as cartoon characters. The difficult is easy and the easy difficult for that tiny brain, or so it seems.

Hearing similarly involves complexity and simplicity. The only physical event on which hearing is based is the variation of air pressure at the ears. A plot of the sound pressure generated by a person saying "science" in a quiet room is completely different from the plot of that same word said by the same person against a background of white noise, such as cocktail party chatter. A person has no difficulty hearing the message in the noise, although the physical sound patterns are quite different. The ability of the human ear and brain to recognize the word in the noise is vastly superior to that of any machine.

Understanding such abilities is the current focus of research on perception. This research includes innovative theoretical work and sophisticated experiments with animals, and it is increasingly able to use computers to develop theoretical models and simulate experimental tests. The research involves psychologists, biologists, physiologists, physicians, graphic artists, physicists, engineers, and computer scientists. The work is highly interactive: for example, behavioral analyses of sensory and perceptual processes provide evidence for functional distinctions that can then be sought in the workings of the brain. Another example comes from theoretical advances in the development of recursive computer models that capture the simultaneous use of multiple sources of information. These are often called activation models because each piece of information that plays a part in a particular mental process is viewed as an active influence on the direction of the process. These activation models are being applied to previously intractable problems in research on human and animal vision.

Work with various species of animals commonly used in laboratory experiments is playing an important role in learning how brains accomplish the tasks of seeing and hearing. By recording the electrical behavior of single nerve cells and studying the wiring diagram of how these cells are arranged in the brain, researchers are beginning to understand the architecture of perceptual systems and how they process information. In addition to monkeys, animals such as cats, rabbits, goldfish, and barn owls have yielded valuable information about sensory systems. Which species is appropriate depends on the particular ques-

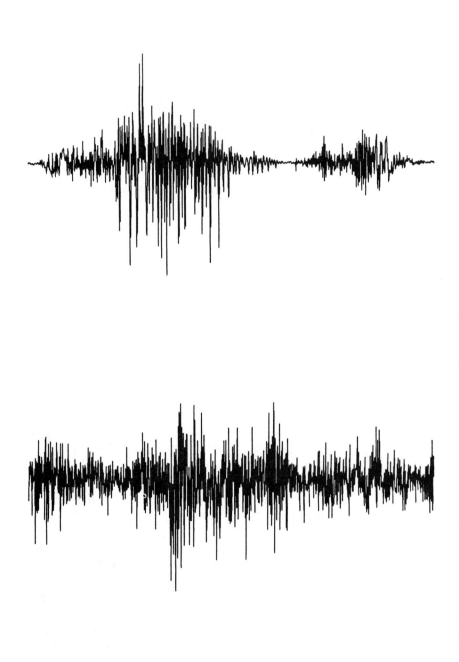

tion being asked and the similarity of that aspect of its perceptual system to the human system.

Visual Analyzers

The retina of the eye—really a bit of exposed brain—is a regular mosaic of five different layers of nerve cells whose spatial arrangement affects the nature of vision. Aided by recent technical advances in electronics, researchers have shown that in one early stage of visual processing, millions of small groups of neurons in the eye and the rest of the brain work as parallel microprocessors or analyzers. Each analyzer responds preferentially to stimulus components within a narrow range of size and orientation; they decompose the scene into a vast number of overlapping components. Precise quantitative models based on such analyzers can be explored by complex computer simulations and can now explain how the human eye and brain detect and identify low-contrast visual patterns, such as nighttime shadows. Some current work suggests that the brain may be able to rapidly select and regroup the analyzers, depending on what is being perceived.

One application of such models is to evaluate, at the design stage, the relative clarity and efficiency of a variety of visual displays: flight simulators, video

---

HEARING   How do the senses, such as hearing, work? How does a person's auditory system identify a single voice in a buzz of conversation and pick words out of a vocal stream?

The figures on the facing page show the sound pressure generated by a human male voice saying the word "science." They are digitized plots of the amplitude of a sound wave over time; the total time is 0.85 seconds; the amplitude is sampled at a rate of 10,000 observations per second.

In the top figure, the person spoke in the stillness of a soundproof chamber. In the bottom figure, he spoke while his voice was masked by noise that simulated a roomful of people talking, such as would occur at a cocktail party.

Neither the human eye nor any known form of sound filter or computer analysis can detect the wave form of the top figure in the bottom figure. That is, they cannot extract the vocal wave form from the noise. But a normal listener has no trouble hearing and understanding the word through the noise. How the human ear and brain perform the analysis needed to recognize the word in the noise is a fascinating and complicated scientific problem.

terminals, warning lights, and information signs. Another application is image augmentation. The inherent inadequacies of electrical and optical imaging systems—such as blurring due to the atmosphere—can be compensated for by computers. Information that is particularly useful to human perception can be improved at the expense of less useful information. These applications are aimed at speeding up such tasks as reading x-ray plates and satellite reconnaissance displays, tracing circuit diagrams, and following blueprints and maps, which are done slowly and are error-prone without augmentation. Another application in the future will be the possibility of more efficient transmission of visual images over telephone lines.

Researchers have discovered that some visual tasks are so complex that humans, some primates, and possibly many other animal species use special-purpose brain areas to solve them efficiently. For example, the human ability to recognize and discriminate readily among a seemingly endless variety of faces is now known to make use of certain neural circuits in the right-posterior cerebral hemisphere. Facial recognition may be different from the recognition of most other kinds of objects in the world, but it may be that other forms of expertise in pattern recognition hinge at least in part on engaging those neural circuits. Researchers are beginning to explore in detail the extent and nature of such special-purpose systems in the human and animal brain and their relation to perception.

Color Constancy

An outstanding example of a very important special feature of visual perception is color constancy: color relations between objects in a scene seem relatively invariant to the eye, independent of the light playing on the scene. A blue shirt looks blue whether seen in sunlight or in a dimly lit restaurant. Yet this invariance cannot be replicated by color photographic media.

Recently, some investigators have proposed a theory about how the eye and brain may separate the inherent color relations among objects from the light impinging on them. Certain newly defined mathematical algorithms, if carried out simultaneously over the entire visual field, presumably by millions of visual analyzers, have been shown in theory to be able to separate the inherent colors of objects from incident lighting. This theoretical model assumes certain mathematical conditions that correspond to qualities of colors, of light, and of the high-level "programming" of the brain. Large-scale computations are now being carried out to test and refine the model and to find ways to simulate it in real time with new computer architectures.

If, as the theory suggests, this decomposition works efficiently and if, as is anticipated, the electro-optical technology for recording a visual scene digitally can be made sufficiently compact, a new form of photography could result. Instead of the various "speeds" of film suited to different lighting conditions, one would need only one type of photosensitive medium that could be used under any circumstances. The "developing" process would then decompose what is inherent in the scene from the lighting that happened to exist at the time of the photograph. With such a process, one could display or print the picture with whatever lighting conditions one wanted. The developing and printing processes, carried out on a computer, would be analogous to what the brain does all the time. The ability to adjust lighting as if from the mind's eye would mean unprecedented expansion of the capabilities for the scientific study of vision and for applications in graphic arts, industrial design, and computer vision.

Depth and Motion

That the world is spatially three-dimensional while the retina is spatially two-dimensional creates inherent visual ambiguities. The eye cannot directly capture motion perpendicular to it; the eye and the rest of the brain must infer depth and motion from ambiguous clues. Resolving these ambiguities, which is usually essential for accurate vision, is possible because of certain constraints that occur in natural scenes concerning shading, motion, texture, and so on. Recent results of behavioral experiments with humans and animals, which have inspired corroborating neurophysiological experiments, reveal that higher-level neural analyzers of motion in visual images are sensitive to the global direction of a pattern's motion, even when parts of the pattern are moving in other directions. These analyzers give rise to moving optical illusions, such as rigid objects that are perceived as flexible and vice versa. Another example is the illusion that very large objects are moving more slowly than they actually are. Landing or taking off, a jumbo jetliner appears to be flying more slowly, by a sizable factor, than a small, executive jet, when in fact their speeds are very similar. This kind of size-speed illusion is believed to be the underlying cause of many railroad-crossing accidents when motorists drastically underestimate the speed of an approaching train. These illusions, which are difficult and in some cases virtually impossible to study experimentally except with modern computer-graphic technology, imply that immense distortion in visual displays may be tolerated or even go unnoticed by both humans and animals.

Perceiving Objects

The process by which visual analysis fills in a big picture from a collection of details means that mere fragments of objects suffice in normally cluttered scenes to permit observers to produce coherent, conceptually appropriate perceptions. Adults make use of several experimentally confirmed theoretical principles to do this, such as assuming that objects in a scene do not share boundaries. People also tend to see correlated movement of disconnected parts in natural scenes as a single moving object partially masked by another object. This latter principle is apparently a deep-seated tendency and has even been demonstrated in very young infants. Increasing knowledge about visual perception can be expected to improve the design of factory robots that must make complex identifications of objects in order to carry out their function without selecting the wrong object or inadvertently hurting someone.

Temporal Auditory Patterns

Hearing problems having to do with space (for example, distance and localization) are very important in the design of sound equipment and auditoriums and have long been studied. Much of the current focus of research on auditory perception, however, concerns complex patterns of sound stimuli varying in time, such as speech and music. Auditory signals can now be designed and stimulated quite precisely by a computer driving a digital-to-analog signal coverter, and these artificial sound patterns are used to study specific aspects of the hearing process.

One type of study measures a listener's ability to discriminate changes in the intensity of one of several tones that are played simultaneously. Initially, the ability to sense a change in intensity within a complex of other tones is far less than when a single tone is played by itself. But with practice in listening to complex tones repeated with little variation, the ability to recognize the variations when they do occur becomes very good. Studies are now under way to see whether practice in listening to complex sound patterns is characteristic of how an infant learns to identify the particular phonemic differences that characterize its prospective language.

An important area of application of research on the perception of auditory patterns (including speech recognition, described more fully below) is in the design of hearing aids. They have been greatly improved in the past decade by coupling what was known about the nature of hearing loss with modern electronics. Whenever the auditory deterioration is peripheral rather than central, certain aspects of the loss, such as reduced loudness at certain frequencies, can be mitigated by appropriately altering the sound waves at the eardrum. Whether it will also prove possible to compensate for the inability to separate speech signals from background noise may depend similarly on the nature of the loss, which further research is likely to reveal.

Music is the focus of much scientific research. Musical keys and scales are highly structured, and musicians, mathematicians, and philosophers have long speculated about the underlying nature of that structure. More recently, the rules of counterpoint and orchestration are being explained, not as arbitrary requirements of particular musical styles, but as selective adaptations to basic tendencies of the human auditory system to group sounds in certain ways as a step in auditory pattern recognition. Researchers some time ago postulated that a set of ratios corresponding to a helical geometric structure underlies tonal perception. Recently, far more systematic studies, using trained musicians as respondents and multidimensional scaling methods (discussed in Chapter 5) to analyze the results, have led to a modified understanding of the geometric relations implicit in tonal perception; they appear to be conical rather than helical configurations. Stimulated by the success of linguists in analyzing language as a rule-governed system, behavioral researchers working with musicians have also successfully begun to define the deeply embedded rules that become internalized in the course of growing up in a musical culture, and they have demonstrated how such deep rule structures strongly affect listeners' perception and memory of what is heard.

MEMORY

Most people can readily find their way around a once frequented neighborhood, improve a sporting skill with practice, or recall an old story. These deeds are possible because the capacity for memory storage is vast; indeed, virtually everything discussed in this chapter depends critically on humans having very large, readily accessible banks of organized information in the brain. Not only is the capacity large, but also much of its use is automatic: learning and memory often occur incidentally, with little special effort. Yet for some people memory failure can become an acute problem. In certain diseases, such as Alzheimer's, the ability to learn and remember is drastically impaired, and life becomes a series of unconnected moments.

Some questions about memory are of long standing: How are memories organized? How does the brain code, store, and retrieve them? What are the computations and processes involved? What happens when the capacity for memory becomes impaired? How can memories be improved? For both these questions and newer ones, a whole new approach to the study of memory derives from the development of computers. However, although much has been learned from the way information is stored in computational systems, the analogy between computer memory and human memory is imperfect in many significant ways.

In the past 20 years, an interdisciplinary revolution has occurred in understanding the organization of learning and memory and their biological foundations. Behavioral studies in animals and humans are characterizing the categories and properties of learning and memory; research on human memory and the brain is identifying the neuronal systems that serve different categories of memory; memory trace circuits in the mammalian brain are being defined and localized in animal models; researchers are beginning to understand the neuronal, neurochemical, molecular, and biophysical substrates of memory in both invertebrates and vertebrates; and theoretical mathematical analysis of basic associative learning and of neuronal networks is proceeding rapidly. Better mathematical and computational modeling of elaborate memory processes, carried out to simulate aspects of specific brain architecture, has become feasible because of the "computer revolution." The newest phase of that revolution—the availability of massive, parallel computer architectures, in which many specialized unit devices process pieces of information simultaneously, seemingly analogous to the way in which neural memories operate—will almost certainly facilitate future work on memory.

Types of Memory

Of the many questions about memory, one of the most fundamental concerns the types or varieties of memory. In recent years, both psychological and neurobiological work have suggested that memory is dissociated into processes or systems that are fundamentally different. For example, amnesic patients with brain injury or disease exhibit severe inabilities to recall and recognize recent events and have difficulty learning new facts or other kinds of information. But, these patients possess some relatively intact learning and memory abilities: for example, on tasks such as manual-dexterity learning trials, they perform as well as healthy and uninjured people, even though they may have no conscious memory of having performed the task before. This evidence—that some kinds of learning can proceed normally even when the brain structures that mediate conscious remembering are damaged—supports the general proposition that there are distinct, dissociated types of memory.

Analysis of recall performance shows that memory is an active process of seeking and reconstructing information, not a passive recording and reproducing of events. Thus, expectations of what things should look like or the way events should happen influence what people notice and remember. For example, after listening to a story presented in jumbled order, people still tend to remember it as being told in proper sequence, following certain widely accepted schemas for what constitutes a story. People also tend to pay little attention to the details of routine situations; consequently, people often remember that the most probable things happened even when they did not. This phenomenon has been demonstrated clearly in the context of eyewitness court testimony. And because people tend to remember events and places in terms of expectations and general knowledge of the world, experiments have shown that memory of an event can be modified or distorted by the manner in which questions about the event are posed. Such experimental findings have important theoretical implications for understanding the formal structure of memory, and they have practical implications for legal proceedings (see "Small Groups and Behavior" in Chapter 2).

Behavioral studies in animals have also revealed certain persistent constraints on learning and memory. Many species, ranging from invertebrates to primates, can learn very quickly to associate or connect distinctive tastes with subsequent nausea, but they are much slower to associate those tastes with prompt threats of pain. In contrast, certain visual and auditory signals are readily associated with pain episodes, but are not so quickly associated with subsequent nausea. These findings also support the general proposition of different types of memory.

Animal studies have illuminated other important aspects of natural learning. In foraging, for example, animals search efficiently among many possible food sources, using past experience to reckon the best trade-offs between relative distances (and associated caloric costs), probability of finding food, and relative nutritional quality (especially caloric content) available at alternative sites (see "Patterns of Food Consumption" in Chapter 2).

Imprinting provides an example of biological adaptive mechanisms for natural learning. Song learning in birds that have a species-typical song is perhaps the most striking example of biological preparedness to learn. Young birds apparently have a "song template" encoded in the brain so that they best learn the species-typical song during a critical period early in life. Some birds, like the canary, learn a new song each year. This capacity for annual song learning has recently been found to depend on an extraordinary biological mechanism. Neurons in the "song circuits" of the brain grow during a period of song learning, extending their connections to other cells and increasing the opportunity for interaction among neurons, and then shrink at the end of the period. These findings may be relevant to examples of human learning that appear to operate under strong biological constraints and occur best during early years— like language learning.

Brain Structure and Neurotransmitters

Memory is under intense investigation at the neurobiological level as well as at the behavioral level. Recently developed techniques for inducing amnesia in nonhuman primates offer great promise for understanding the neural circuits underlying particular aspects of memory. Surgical brain lesions in monkeys, involving the hippocampus and related structures, produce the same selective impairment in memory as that which occurs in human patients who have experienced comparable cerebral damage due to head injury or stroke. Performance is poor in memory tests that require stimuli to be recognized as familiar or after a long delay, but skill learning is intact. This research will lead to identification of the precise set of brain structures and connections that, when damaged, causes human amnesia. At the same time, parallel work on other mammals, especially rats, will be useful in identifying the neural systems involved in acquiring and storing different kinds of memory. This work should make it possible to specify clearly what these systems' functions are and to investigate how these systems work at the cellular/neurophysiological level.

Neural accounts of learning depend heavily on and benefit importantly from the highly developed study of learning at a behavioral level. Once memory circuits are defined in invertebrate and vertebrate nervous systems, their performance capabilities have to be measured. How neural circuits are organized is a mathematical problem of enormous complexity that can only be solved by mathematical and computational modeling. One such complex network model, an approximation to visual cortex, was shown a few years ago to be computationally in agreement with a large number of experimental results.

Recent, unexpectedly rapid progress has been made in identifying essential memory trace circuits that code, store, and retrieve associative memory in the brains of birds and mammals. The kinds of memory under study include the learning of discrete, adaptive behavioral responses and the conditioning of involuntary responses. For one well-studied type of associative learning, classical conditioning of the eye-blink response, the essential neural circuitry required for the reflex has been partly identified. Moreover, locations have been found within that circuitry where memory traces are likely to be stored when conditioning of the reflex occurs. In this case, there is growing evidence that at least one site where memory traces are stored is the cerebellum, and neurophysiological research is being carried out on the interpositus nucleus where changes related to learning can be studied within a volume of one cubic millimeter. Most of this work involves pigeons, rabbits, and baboons.

Other active research is clarifying the role that the brain's neurotransmitters and neuropeptides play in modulating memory. Studies with drugs and experimental animals show that memory can be amplified or diminished by specific pharmacological treatments. Full identification of the core-memory circuits for basic associative learning in mammals is very likely in the next few years. As these circuits are identified and the storage sites located, scientists can make substantial progress in analyzing the detailed storage mechanisms, particularly those underlying long-term and permanent memory. These basic cellular and molecular mechanisms are now the focus of great interest and excitement. For example, in a well-studied invertebrate, the sea hare, simple, biologically universal forms of learning like habituation and sensitization are being related to changes in the readiness of particular neurons to release a transmitter. From work in both invertebrate and mammalian preparations, the general picture emerging is that learning involves the reduction of one or more kinds of potassium conductance through the membranes of nerve cells. Such a reduction in ion conductance may constitute an essential step in how nerve cells change in response to input, thereby profoundly altering the way those cells receive and transmit information. This process is thought to be brought about by phosphorylation of specific substrate proteins, mediated by "second messengers," which can give rise to long-lasting neuronal changes. Second messengers are attractive molecules because they themselves can serve as the agent for short-term memory, while their intracellular effects can include the genomic regulation involved in the cellular changes that occur in the foundation of long-term memory.

The cellular and molecular mechanisms emerging as important in animal-model systems appear to be recurrent in the evolution of species, offering the promise that these models may be of general significance to the question of how human memory is formed and retrieved, contributing ultimately to greater understanding about the functioning of the human mind and brain. Work is proceeding at all levels from abstract mathematics to exploration of molecular and biophysical substrates. Continuing advances in such work may lead to significant applications in the treatment of memory disorders, such as those associated with amnesia, seizures, stroke, and Alzheimer's disease.

COGNITION AND ACTION

New questions about the mind have emerged in this century: How do people acquire knowledge, use it in reasoning, and turn it into practical action? How do human beings and other animals construct mental categories in a world of particular objects and episodes? How do people manipulate images in their minds, train their fingers to perform delicate handiwork, or judge probabilities and decide which risks to take?

Recent discoveries regarding these mysteries have resulted from new scientific methods for observing, describing, replicating, and analyzing cognitive structures and processes. To study human infant abilities, for example, researchers exploit the finding that infants look preferentially at novel objects and events, work at sucking or turn their heads to hear a sound or bring a picture into focus, and give characteristic responses to changes in the environment. For studies of adult attention, researchers take advantage of small but consistent differences in the time-course of mental events. To map the rules that guide behavior, researchers rely on the human ability to detect errors in rule application.

Neuroscience data suggest that complex behavior and mental activity emerge from simultaneous and parallel contributions of many specialized component parts. One possibility is that the basic processing units are neurological "columns" or "modules," each containing 1,000 or fewer nerve cells; if this is true, then there are more than 10 million such functional units in the human brain. A major challenge for future study is to identify more fully the functional units of organization in the cortex, the most highly evolved and complex part of the primate brain. This effort, which is already under way, involves neuroanatomy and physiology together with the analysis of cognition and perception.

Studies of complex intellectual task management take advantage of a wide variety of methods, such as computer simulations. With the increasing availability of powerful computers oriented toward symbolic processing, theories can be tested in the form of programs that simulate the hypothetical mental activities involved in understanding and reasoning tasks. Similarly, mathematical, logical, and other formal models serve to make hypotheses about mental structures explicit and precise. Empirical methods for testing such hypotheses use detailed observations, including protocols that have individuals think aloud while they work, complex tracking systems that record and analyze eye fixations while a person is examining visual information, and repeated runs of computer models to see how well they perform in processing a variety of materials, such as texts to be analyzed or symptoms to be diagnosed.

Early Cognitive Development and Learning

Studies of early human development give some of the strongest examples of how new observational methods have led to new knowledge. The once-common belief that the experiential world of an infant is mostly sensory chaos has been displaced as evidence has accumulated showing that infants are especially sensitive to those subtle distinctions among oral sounds that are important in learning to speak, to cues of visual depth and distance, and to other precursors of complex perception.

The techniques used to show these facts make use of an infant's innate curiosity and tendency to prefer novel sights and sounds over very familiar ones. Experiments built around these techniques are now able to investigate infants' ability to form abstract concepts. For example, infants as young as 6 months are able to match the number of items they see in a display with the number of drumbeats they hear. In a series of studies, infants were sat either on their mother's lap or in an infant seat and shown pairs of slides, one displaying three objects, the other two objects. The pictures of familiar objects used included a comb, apple, scissors, crayon, and book. The objects in the slides varied from trial to trial, and on each trial the infants heard either two or three drumbeats emanating from a hidden, centrally placed speaker. In one set of experiments, infants consistently tended to look at the visual display in which the number of items matched the number of drumbeats played. In experiments similar in design but with significant changes in conditions such as exposure time, infants consistently preferred to look at the display that differed in number from the number of sounds they heard on a trial. Thus, infants can respond to numerical information, in terms of whether pairs of stimuli are equal or different in number. How they do this and whether they use adultlike numerical processes is the subject of much current research.

This line of study at first evolved an elegant theory that young children progress through relatively well-defined stages of understanding in which fundamental structures not present at one stage come into being virtually discontinuously at a later age. Over time, the theory has changed as new evidence has led to new hypotheses and more sensitive methods to testing them have been developed. Evidence now exists that even 3-year-old children can use general principles to understand the structure of stories, organize learning about cause and effect, represent numbers and space, analyze and reason about event sequences, and rapidly acquire the meanings of words. These findings have provided a basis for fundamental reconceptualizations of early childhood learning, which now appears to be enabled and constrained by available mental structures and fostered by a pronounced tendency of the young to seek out particular environments that challenge and nourish cognitive development. Instead of leaps across cognitive gaps, development involves the gradual accumulation of insights and adjustments that cumulatively constitute profound increases in cognitive capability. Current efforts to spell out the mechanisms of such learning benefit from detailed descriptions of development within particular domains of knowledge as well as formal innovations (for example, theories describing the conditions that make a language learnable), computer simulations of change, and the investigation of social-environmental conditions that foster or interfere with self-motivated learning, knowledge transfer, and problem solving. (The case of language learning is discussed later in this chapter.)

Categorical Knowledge and Representation

One aspect of development is the creation of categorical knowledge—concepts and relations among them—which is the preeminent form of mental framework for interpreting and understanding experience. How are categorical ideas represented in human knowledge? A traditional view holds that categories are defined by a list of necessary and sufficient properties that an object must have in order to be a member. Scientific categories (for example, mammal) often have this characteristic. However, categories in natural language generally do not. This is true of relatively concrete ideas such as dogs, trees, and chairs; more so for somewhat more abstract categories such as animals, fruit, and furniture; and completely for thoroughly abstract concepts such as freedom, happiness, and justice.

Empirical studies point to the possibility that categorical ideas are stored in several different ways. Some appear to be hierarchically organized, allowing people to infer that whatever is true of a superordinate concept is true of those of a subordinate level. For example, whatever an "achida" may be, if it is an animal, it must eat and breathe. Other categories are partly organized around spatial relationships: in order to form a forest, trees must be physically close to each other. Still others are represented in memory as a collection of exemplars, and a new object can be classed as belonging or not belonging to the category according to how well its features match those of remembered examples. People sometimes also form a prototype or a schematic representation by a process of abstraction, and a new object is classified by judging its similarity to the prototype or its goodness-of-fit into the schema. A designer seeking the right "look" or a judge seeking to determine the equities in a tort case may use

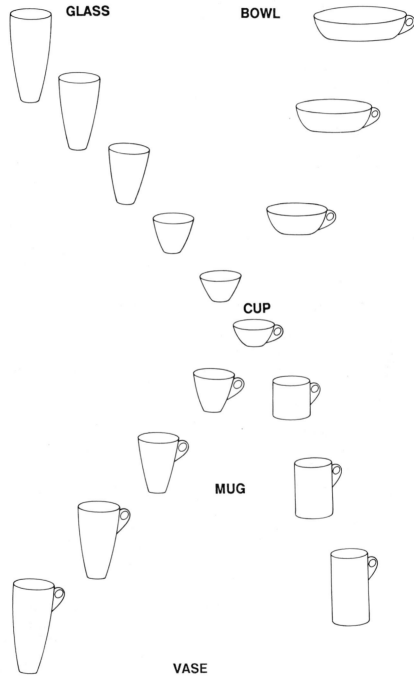

this kind of thinking. Experiments have shown that exemplar-based concepts, schematic or prototypical representations, and associations of features with conceptual categories (including probabilistic associations) are all forms of human knowledge processing, used in different combinations, in different tasks, and by different individuals. Investigators do not yet know what principles govern conceptual organization in human beings (or other animals); discovering those principles is a key to understanding many human mental functions, including speech perception, the organization of memory, and learning, and it will also contribute to a variety of computer applications to problem solving.

Imagery

Not all human knowledge consists of propositions, concepts, and principles that can be expressed verbally. Crucial components of knowledge are based on visual and auditory imagery and on patterns of motor activity. Research during the past 15 years has made major advances in understanding the properties of spatial cognition and action and ways in which brain structures and processes are involved in cognitive functioning in different domains.

Major advances in the study of visual imagery have included identification of specific mental operations for manipulating spatial information. A landmark

---

CATEGORIZATION   How are categories defined in the mind? How do people recognize instances of categories? These questions are fundamental to understanding human thought, since reasoning about things is based on knowledge about them, and knowledge about things is largely knowledge about categories of things.

Sharp boundaries between categories, based on strictly defined sets of properties that an object must have to belong to the category, occur in many scientific concepts but rarely in the world of experience that is expressed in natural language. For example, what makes something a "cup"? The evidence of numerous experiments in several languages is that it helps for the object to have a handle, to have sides that bow inward, to contain a beverage, to be made of ceramic, and to look slightly wider than deep. A tall glass cylinder with no handles and a flower in it is clearly not a cup, but something can still be a cup if it has straight sides, or if it lacks a handle, or if it has a flower in it—as long as it has most of the other properties associated with cups. As particular properties change, an object tends to be recategorized into distinctly different categories such as glass, bowl, or vase, sometimes passing through transitional or subordinate concepts such as mug. This figure shows such categories and characteristics.

From a wealth of such findings arose the idea that, in natural language, category membership is a graded function of the typicality of the properties of an object relative to other members of the category; a category is thus represented in the mind by an abstract prototype and specifically remembered examples, and a new object is compared to both.

demonstration based on a typical performance IQ test item established that the decision as to whether two simple figures are the same or different is based on an imaginary mental rotation of the image, much as one would physically rotate an object with one's hands. The time required for such judgments has been shown to depend approximately linearly on the angle through which a figure has to be rotated in the mind's eye. This finding is consistent with the hypothesis that such mental operations are analogous to spatial ones. Other tests about the nature of spatial reasoning draw on quite different spatial operations, such as expanding or compressing the size of an image or inferring the appearance of a scene from different perspectives. These tests are also consistent with the hypothesis of some sort of analog representation, although in each case one can also devise explanations in terms of descriptive statements about the object.

Not all physical properties of objects need be encompassed by the cognitive mechanisms of imagery. Researchers are now working on such questions as: Which properties of physical transformations are preserved in mental images? What are the differences in imagery skills among individuals? How do these differences arise? It is now established that imagery is not a unitary, undifferentiated phenomenon but consists of distinct spatial abilities. Individuals can be good or bad at a given component of imagery ability independent of the other components. Research on imagery is leading toward an increased understanding of the cognitive deficits that can result from brain damage, as well as useful characterizations of intellectual differences that may be relevant to observed male-female differences in intellectual style, science and mathematics education, and other scientific and practical questions.

As the imagery example illustrates, one legacy of research on intelligence testing is the idea that there are different kinds of intelligence. Some theorists are exploring the related hypothesis that there are specialized modules of thoughts, each with separate sets of principles, not only in language but in other forms of symbolic processing, such as in the way all living creatures learn to find their way in space. It is striking but taken for granted that people can recognize places while apparently lacking detailed information about them. For example, many people can retrace a route along city streets, knowing when to turn at a particular corner, without knowing street names or being able to recall in detail what is at that or any nearby corners. What does this mean and how does it relate to imagery? To what extent could things be scrambled or replaced and a location still be recognized as the same place? How can robots be programmed to recognize readily where they are? The problem of recognition is clearly important and most likely complicated, even though many animals seem to possess the skill. Work on this topic is pursued fruitfully by researchers at many levels of analysis, from neurophysiologists mapping the locations of these functions in the brain to psychologists, linguists, anthropologists, city planners, and ethologists describing and modeling the behaviors involved in moving efficiently and communicating efficiently about location.

Individual Decision Making

Decision making—choosing among options—is a central topic of research on modern life. Studies of the decision making of individuals have followed two broad, intertwining pathways—normative and descriptive. Research in a normative vein seeks to define the conditions, practices, and procedures under which decision makers can achieve some prescribed goal, such as maximizing expected utility or satisfaction, achieving economic efficiency, or securing democratic representativeness. A central line in normative research is to illuminate rational selection among alternative actions that have complex patterns of possible outcomes depending in part on chance and uncertainty in the environment. Research in a descriptive vein seeks to understand the mechanisms and procedures actually followed by individuals in reaching decisions, particularly when normative approach is difficult to specify or calculate. Each mode of research stimulates the other, and many assumptions, hypotheses, and findings are common to both.

Normative research has traditionally been guided by a theory based, in part, on the principles of mathematical probability and, in part, on calculation of trade-offs between the values of outcomes and the probabilities of their actual occurrences. The modern development of normative decision theory began in the 1940s with the introduction of an elegant mathematical theory of games and economic behavior that laid out a rational basis for making choices among actions whose outcomes are partially determined by chance events with known probabilities of occurring. This theory reduces in practice to a rather simple numerical model for computing which alternative move in a situation has the largest average utility and should therefore be the one selected. Despite its logically compelling character, however, the theory does not seem to be followed by most people when they make individual decisions. As a result, a noteworthy line of experimental study is trying to formulate and systematically understand in what ways people depart from the normative theory.

A considerable body of economic and statistical work has been built on normative decision theory, and it has proved powerful when applied to behavior in various financial and insurance markets. But objections have been raised against applying the theory in other areas, both by decision theorists on formal grounds and by experimentalists carrying out carefully designed studies of the choices people make under well-controlled conditions. In the past 10 years, the empirical results consistently showed departures from rationality postulates, and these findings have created a new challenge for theory development. Because the empirical findings are complicated, it is as yet unclear exactly which one (or more) of the basic postulates of rationality is the principal culprit and needs to be modified. Further experimental studies and development of more satisfactory theoretical fundamentals are proceeding.

A study by physicians and behavioral scientists of how people use statistical information in decision making highlighted the importance of how the possible outcomes are perceived. More than 1,000 people—graduate students in business school, physicians, and medical patients—were asked to imagine that they were suffering from lung cancer (none was known to have this disease) and could elect one of two treatments: surgery or radiation. The subjects were presented with the statistical prognosis for each treatment: for one group, the outcome was stated in terms of the probability of surviving for various lengths of time (for example, a two-thirds chance of living at least 1 year following treatment); for the other group, the outcome was stated in terms of the complementary probabilities of dying within those lengths of time (for example, a one-third chance of dying during the first post-treatment year).

When presented with the two therapeutic options framed in terms of survival chances, the students chose radiation over surgery only 17 percent of the time, while the students presented with the identical prognoses framed in terms of the chances of dying chose radiation 43 percent of the time. When radiologists were the subjects, the contrast was just as pronounced: 16 percent chose radiation when presented the survival prognosis and 50 percent chose it when presented the mortality prognosis. Among patients with chronic medical conditions, a considerably older group, the results were 20 percent and 40 percent in preference for radiation. Since the situations described to the subjects were simple and logically identical, these data clearly contradict a postulate of traditional decision theory that identical situations be treated the same. The framing of the problem clearly matters. Various studies are now in progress to expand and test several alternative theories of these framing effects, and research in the next several years is likely to bring major developments in this fundamental area of inquiry.

A major line of work is focused on the fact that individuals often appear to rely on conventional biases, simplified concepts, and common rules of thumb in making decisions, rather than developing close probabilistic calculations or estimates. Although such heuristics typically reduce informational and cognitive demands for reaching decisions, they lead to demonstrably and systematically incorrect results in certain very common situations. For example, the heuristic of representativeness leads people to think that a more "typical" event is also a more probable one, even when this is logically impossible. In one experiment, respondents were asked to report on the relative likelihood of various possible events, an example being the performance of championship tennis player Björn Borg in a hypothetical Wimbledon match. A substantial fraction of subjects thought it more likely that (a) Borg would lose the opening set but still win the match than that (b) he would lose the opening set, whatever the final outcome. Yet possibility b has a likelihood equal to or greater than a, since b includes every instance of a as well as the additional possibilities of losing both the first set and the match. This kind of "cognitive illusion," attributed to people using the heuristic of representativeness, has been confirmed in many experiments.

Two other common heuristics are availability and anchoring. Availability refers to the fact that estimates of the likelihood of an outcome are unduly influenced by the ease with which examples of particular outcomes are brought to mind. For example, most people will conclude that there are more words with r as the first letter than the third because it is easier to build a list of the former, but the latter are in fact more numerous. In anchoring, an individual's final estimate or judgment of a situation is overly influenced by the first of multiple examples or reference points that are observed; for example, extra weight is attached to the first of a series of numbers whose average (mean) is to be estimated. Knowledge about these and other heuristics and framing effects is laying the foundation for a more general theory of how information is used by individuals, and consequently, for a more thorough and precise understanding of individual decision-making behavior in general.

Framing effects are recognized and manipulated on an intuitive level by publicists, advertising specialists, and politicians, among others. However, new formal models incorporating framing effects and decision-making heuristics are likely to have important new policy applications. For example, studies of framing effects may affect how ingredient or warning labels are written, how truth-in-lending laws are formulated, how unit prices are displayed in supermarkets, and how election ballots are designed.

Progress in understanding decision making has relied heavily on theoretical analysis and questionnaire-based experiments; there have been only limited observations of behavior in real decision situations. Now, however, researchers are beginning to use more complicated and realistic methods of dynamic study, including interactive computer-run experiments, improved field observation methods, and refined techniques of statistical inference to study actual behavior. A unified understanding of framing, biases, and heuristics, including the conditions that generate them, their robustness and their consequences, will become more possible as researchers are able to characterize choice behavior as a multistage process that involves information, evaluation, expression, and feedback. A promising parallel development is the fuller inclusion of these characteristics of choice behavior in systems models, working out their implications for market efficiency and other aspects of organizational performance.

Reasoning, Expertise, and Scientific Education

How can general science education provide more students with durable scientific concepts and principles? It is widely recognized that such instruction often fails to communicate the fundamental meaning of scientific concepts. Indeed, recent research has dramatically shown that many people interpret physical phenomena in ways that are contrary to principles that they have apparently learned well, at least to the extent of being able correctly to solve typical textbook problems. For example, many students who know how to calculate the Newtonian formulas fail to invoke the Newtonian principle of inertia when asked to sketch roughly the path followed by an object dropped from an airplane in flight. Researchers ask: Do people base their incorrect judgments on fragments of knowledge that come from experience, such as the way in which objects ordinarily fall when they are dropped? Or, do people have relatively coherent, but incorrect, naive theories about motion, gravitational force, and the like? Since people cannot report how their judgments in these matters are formed, current researchers are attempting to find out through indirect approaches. The answers have important implications for science education because one uses quite a different set of instructional methods to teach students when to apply conceptual schemes than one uses to bring about major reconstructions of the students' naive theories.

The handling of concepts, reasoning, and skills by "experts" is of great interest to information technologists and educators. Initially, two complementary ideas about experts were prevalent: that they have exceptional mental capacities, such as an innate or trained ability to retrieve more facts or consider more possibilities at a time than do nonexperts; and that they have accumulated more knowledge about their subject than nonexperts, often because they have many years of experience. These ideas were incorporated into the design of early programs in artificial intelligence, in particular those for chess, for which the main goal was believed to be an ability to selectively consider many moves in advance. These ideas also underlie the design of contemporary computational "expert systems," most of which incorporate large collections of knowledge. These same ideas are also reflected in many features of the educational system, where achievement in teaching and learning is often assessed exclusively with tests of factual knowledge or computational accuracy.

However, researchers have shown that the underlying structure of knowledge is at least as important as the amount of information. Although experts do remember a great deal of specific information, their capacity to do this mainly depends on their having acquired elaborate, highly organized structures of knowledge. For example, chess experts are hardly better than nonplayers at remembering a randomly jumbled set of pieces on a chess board—but they are far better at remembering a coherent board. Furthermore, experts, like novices, are typically not fully aware of the principles by which their knowledge is organized and used, so these tacit forms of knowledge are not reported by them. Thus, most expert systems now in use neglect some of the most important aspects of expert knowledge.

One such well-known system was designed to assist in the diagnosis of infectious diseases and prescription of antibiotics. Initially, the knowledge simulated in the system was a set of relatively simple rules of inference and a program that evaluated hypotheses simply by accumulating the positive and negative evidence it received. It matched reasonably well the judgments of successful physicians about the specific knowledge they used in their diagnostic work. But when the system was extended to aid in the training of physicians, it did not work well; it became evident that the organization of its knowledge and its methods of reasoning were seriously deficient. The program was reorganized based on more thorough analyses of the knowledge and real-time reasoning sequences of physicians in their diagnostic and training activities. These analyses showed that diagnostic strategies, including the use of hypotheses in selecting questions in interviews and the comparison of symptom patterns with the physician's mental representation of disease conditions, play a crucial role in diagnostic performance. These usually tacit components of knowledge have to be explicitly considered in physicians' training; the latest version of the expert system attempts to bring such tacit knowledge into play.

The importance of general concepts and principles in expert knowledge is also demonstrated in recent research on problem solving in physics, in which expert and novice performance has been contrasted. An expert's understanding of problems includes the general qualitative concepts and principles that the problem illustrates, such as conservation of energy and laws of force, and these principles are used to organize the expert's reasoning in the problem-solving process, leading to calculations as a final step. In contrast, a novice's understanding of problems mainly involves more superficial features, such as the kinds of objects in the problem. The novice typically seeks solutions by translating the available information directly into formulas that allow the quick calculation of an answer—correct or not. Indeed, as noted above, students' knowledge of formulas is often quite disconnected from their understanding of general principles.

People tend, as they become expert, to reorganize their knowledge of a given domain, and this tendency is not restricted to adults. For example, young children have much implicit knowledge of the difference between animate and inanimate objects. They know that animals move by themselves, that trees cannot have feelings, that dolls lack brains. But they do not assume that all animals breathe, reproduce, and so forth. Neither do they classify plants and animals together. They come to do these mental tasks, around ages 8 to 10, when they reorganize their knowledge about objects in the world, independent of explicit formal instruction, in accord with an intuitive theory of biology. The growing realization that children regularly reorganize their knowledge is of considerable import. These naturally occurring mental processes help uncover the laws of theory construction and concept reorganization that underlie the acquisition of expertise.

Research has begun to clarify some of the ways that understanding of general principles contributes to expert problem solving and reasoning, and it has also begun to show how children develop intuitive understanding of important general concepts and principles. Researchers are just beginning to be able to characterize that understanding in explicit, testable ways. The crucial issue now is constructing a rigorous theory of the cognitive processes that are usually categorized as "intuition," involving qualitative reasoning about quantitative and other abstract concepts. A promising start has been made in formulating such theories, which will enable experimenters to test hypotheses about the properties of this form of expert reasoning that previously eluded systematic study and theoretical analysis.

Complex Action

The miracle of action is the ability of the mind to produce organized acts related to plans of action, perceptions of the world, and motives. People are not born with this ability. Newborn behavior appears to be comprised primarily of random movements and rhythmic stereotypes; over the first year of life they are gradually transformed into voluntarily guided, intentional, purposive actions. The study of such motor skills from infancy to advanced performance is exploding after a long, relatively dormant period. The basic research interest in this area is complemented and invigorated by advances in robotics, by searches for a better fit between people and machines, by problems that occur in the manufacture and use of motor prosthetics, and by medical concern with motor disorders ranging from stuttering to Parkinson's disease. Exciting and promising results have been facilitated by new optical, magnetic, ultrasonic, and x-ray technologies for transducing motion, for storing the massive amounts of data collected, and for analyzing these large data bases, sometimes using artificial intelligence methods.

Researchers are addressing a wide variety of questions, such as: What aspects of movement does the nervous system control? In accomplishing an action, how does the system constrain the many motions that are biomechanically available? What determines the accuracy of reaching? How are limbs coordinated? How is serial order represented and realized in planned movement sequences? How do errors in speech and typing come about?

Consider a person reaching his or her hand toward an object, a motion controlled by an interplay of elbow and shoulder rotations. Biomechanically, the hand can approach the object along any of a vast number of paths; one might expect the actual path and the velocity profile along that path to reflect this complexity of control and to vary with the conditions of movement. Yet in the horizontal plane, the path is essentially a straight line, exhibiting a single-peaked, bell-shaped velocity curve. More generally, it is the overall movement itself, and not the pattern of activity in individual muscles, that is invariant during compound arm movements. Such simplicity suggests planning at the hand level rather than the joint level, with the system generating complex, coordinated joint-angle changes to achieve simple hand trajectories.

Also in the domain of single actions, there is a new understanding of the remarkably general logarithmic trade-off between the speed and spatial accuracy of limb movements. The prevailing theory had been that precision slows a movement because of an increase in the number of visually guided corrective submovements, with each submovement independent of precision. However, current experiments, growing out of a new mathematical theory, show that the trade-off occurs without visual feedback and that submovement speed varies with precision.

How are single actions combined into ordered sequences? Studies of speech and typing during the past few years have revealed advance planning of entire sequences and the hierarchial organization of actions in multiaction units. For example, in rapid utterances in languages such as English, the unit is not the syllable, or the word, but the stress group, a sequence of syllables containing a primary stress, which usually corresponds to a grammatical phrase or clause structure. This supports the traditional hierarchical model of speech production, but it has also led some researchers to posit a network model, in which activation spreads both "up" from the sensors and "down" from the control nodes of the network. Only a hierarchical model can thus far explain facts of slips of the tongue, and new methods of inducing speech errors under laboratory conditions have facilitated research of this topic. In a similar vein, a model of typewriting, based on parallel distributed processing that converts a sequence of discrete symbols into continuous and temporally overlapping movements of fingers and hands, explains many features of timing and errors in the performance of skilled typists, including how a stroke by one finger can be accompanied by movements of other, then irrelevant, fingers to position them more favorably for action two or three strokes later.

This is an active and exciting period in the study of complex human action. New findings are expected to improve person-machine interfaces (for example, instrument panels and keyboards), skill training, implementation of artificial movement systems (prosthetics, manipulators, robots), the diagnosis and treatment of movement disorders, and understanding of other complex skills like talking and walking.

LANGUAGE

To be fluent in a language is to be able to produce and understand an indefinite number of sentences never spoken or heard before. As one component of this ability, every spoken word can be identified by hearers in less than one-third of a second, drawing on the more than 100,000 forms stored in the mental dictionary of a typical monolingual adult (bilingual or multilingual people store hundreds of thousands of word forms). In little more time than it takes to process the sounds themselves, the words are then assembled into meaningful sentences that more or less correctly represent the message intended by the speaker.

This casual miracle of communication is possible in part because the human brain is uniquely suited to acquire and use language. Chimpanzees and gorillas are now widely viewed as having greater nonlinguistic cognitive abilities than previously thought, but they are unable, even with the most intensive human training (in sign language), to learn 1 percent of the vocabulary that is acquired by virtually any 3-year-old human child. Nor can these primates learn even the simplest of the complex grammatical rules known to nearly any 2- or 3-year-old human child.

Since nearly all humans are fluent in at least one language and, except when learning a new one, seldom consider what makes this possible, the complexity of the knowledge underlying the ability to speak and understand, and to read and write (abilities derived from spoken or signed language), is often not fully appreciated. Though the question of language acquisition and use has puzzled philosophers, educators, and scientists throughout modern history, answers to the puzzle have proven elusive until recent advances that rest, in part, on a sophisticated modular conception of language and its relationship to other cognitive faculties.

While it was once commonplace to view the grammatical properties of language as essentially derivative—a by-product of the general cognitive, physiological, and other nonspecific systems underlying human intelligence—new evidence has convinced a growing number of scientists that linguistic capacity (and possibly, the mastery of grammar itself) is best viewed as an autonomous cognitive system, serving other systems but governed by its own set of distinct principles. For example, discourse patterns have been discovered that suggest that speakers can accommodate only one item of new information in a grammatical clause. Moreover, it appears that speakers restrict the appearance of this new item to certain specific grammatical roles within the clause. This and related results have led to proposals of specific models of cognitive resources that both enable communicative processes and limit their scope.

Principles of language performance and processing by no means exhaust the realm of possible knowledge about language, which is perhaps our richest cultural heritage, produced collectively over thousands of years and used for a great variety of social purposes. But this approach has led to entirely new methods of investigation that underlie some of the most important recent discoveries.

Acquisition

The remarkable human facility to acquire language depends on a rich genetic endowment. People are well equipped for fluency in language (speech and gestures), just as birds are especially equipped to acquire and perform the songs of their species. Newborn infants, for example, respond to acoustic distinctions that are systematically used in some human languages—even though not necessarily in the child's own linguistic environment—in a way that is different from their responses to distinctions that are not linguistically significant in any known language. In a study of speech perception in 4-month-old infants, the child sucks rapidly to hear "pa" or "ba." As infants become habituated, the sucking rate drops. When a new stimulus is substituted ("ba" for "pa" or "pa" for "ba"), the infant dishabituates and sucks quickly once again. Similar results have been obtained for 1-month-old infants.

These American infants thus show a sharp boundary in discriminating between the sounds /ba/ and /pa/, a phonemic boundary in the English language. But these same American infants have an analogous sharp discrimination boundary in a prevoicing region that is not a phonemic contrast in English but is in certain other languages, such as Thai. Adults have considerable difficulty making such discriminations when they are not phonemically contrastive (functionally important) in their language. This indicates that certain aspects of phonological sensibility may be "prewired."

Linguistic abilities can also be dissociated developmentally from other cognitive abilities. There are numerous cases of children who have few cognitive skills and virtually no ability to use language in sustained, meaningful communication and yet have extensive mastery of linguistic structure. For example, one severely retarded young woman with a nonverbal IQ of 41 to 44—who lacked almost all number concepts including basic counting principles, drew at a preschool level, and possessed an auditory memory span of three units (for example, syllables such as "two, three, one")—could nonetheless produce syntactically complex sentences like "Last year at school when I first went there three tickets were gave out by a police." In a sentence imitation task she both detected and corrected surface syntactic and morphological errors. But she did not know how many "three tickets" were and was not sure whether "last year" occurred before or after "last week" or "an hour ago."

Conversely, there are cases of children with little grammar but with other verbal abilities. One girl who was physically and socially isolated from the world from approximately age 1 to 14—with no language input during that period—rapidly acquired a large vocabulary following her liberation, but her utterances remained nongrammatical, devoid of morphological endings (for example, past tense or plural markers) or syntactic operations (for example, converting statements into questions). This contrast between word lists and grammatical rules is indicative of different and distinct abilities.

Clinical studies of aphasia have given dramatic confirmation to these new fundamental theories. For example, local damage to certain regions of the left brain do not lead to across-the-board reduction in language ability, but to selective, deep deficits, consistent with the idea of independent grammatical components or modules. Some patients with left-brain damage make many semantic substitutions in reading words: saying "pixie" when asked to read "gnome," "sick" for "ill," "prison" for "jail". Some can read a word like "tortoise" perfectly but cannot say what it means; some speak fluently but with non-sensical content; others speak in telegraphic style, leaving out all the short function words. Clinicians are able to use linguistic phenomena such as these, together with analysis of the known brain damage and previously observed patterns of correction between such symptoms and brain damage discoveries in autopsies, to improve the diagnosis and treatment of aphasia.

The dramatic new technologies of neuroimaging, such as computerized tomography (CT), magnetic resonance (NMR or MRI), and emission tomography using positrons (PET) or single photons (SPET), now make possible the exact delineation of brain structures involved in various language functions. A major stimulus to further progress in this field may take place when imaging technology becomes more widely available for research with a variety of populations. For example, since aphasias occur among speakers of all languages, research on aphasia can help isolate the basic, universal capacities and neural substrates underlying human language. Because languages like English rely heavily on word order to convey information that languages like Russian signal through inflections (suffixes, prefixes), an important question is whether patients with neurologically similar brain lesions (discoverable through imaging) but who speak very different languages will exhibit manifestations of the lesion that seem very different yet correspond to the same underlying abstract functions.

The organization of language mechanisms is also being studied using electrophysiological techniques, such as scalp recordings of event-related brain potentials (ERPs), which measure the electrical fields that arise from coordinated groups of neurons engaged in processing sensory, cognitive, and linguistic information. By studying how ERPs vary in time, it has been possible to differentiate among certain linguistic operations. The various components of ERPs exhibit asymmetries over the left and right sides of the brain and are sensitive to such factors as handedness and mode of language acquisition (spoken versus signed).

Another important new line of research links language acquisition studies with theoretical work in cognitive science and artificial intelligence. One emerging area is the theory of machine inductive inference, which investigates how intelligent systems develop logical models or schemas based on evidence from their environment: for example, the inferring of the grammatical structure of a language based on utterances heard and overheard. This theory provides a framework for systematic comparison of various learning algorithms in terms of their relative strengths, resource requirements, and behavior in various environments. When combined with empirical studies of language acquisition, the theory of machine inductive inference provides constraints on the character of learning strategies implemented by children and reflects on the character of the class of languages that can be acquired. Such studies are important to system builders in artificial intelligence.

Sign Language

Linguistic research on the sign languages of the deaf, particularly American Sign Language (ASL), is only 25-years-old, and it has opened a very important avenue toward a deeper understanding of all language. In spite of its name, ASL is not a signed version of American English, but rather a complete language in itself. It is more closely related to French sign language than to spoken English or British sign language, which is logical because it was first brought to the United States by teachers of the deaf from France. ASL has all the crucial properties common to spoken languages, including highly abstract underlying grammatical and "phonological" principles. The relationship between the form of a sign and its meaning is as arbitrary as that between the sound of a spoken word and its meaning. Sentence formation in ASL is just as rule governed as it is in spoken languages. ASL uses facial and other simultaneous body gestures (for example, lifting of eyebrows) to convey linguistic information similar to the morphological inflections that occur in spoken language. Variations in tone and emphasis convey additional layers of meaning in both spoken and sign languages—in the latter, through the pacing and shaping of the gestures.

Like hearing children with speaking parents, deaf children with signing parents acquire their native language without formal instruction and in similar stages. Brain studies of normal signers and deaf aphasics (patients suffering language loss following left-side brain damage) show that the left cerebral hemisphere is just as dominant for sign language as for spoken language. This finding has been a definitive result in proving that the left-hemisphere specialization in the brain in language acquisition is not due to its capacity for fine auditory analysis, but for language analysis as such.

Grammatical Universals

Work on formal theories of grammar in the past 25 years has considerably sharpened understanding of linguistic universals—principles that are common to all languages. For example, despite the fact that the rules to form passives ("The ball was thrown by John"), questions ("Who threw the ball?" or "The ball was thrown by whom?"), and imperatives ("Throw the ball!") differ markedly from language to language, modern theory and data argue that such constructions are manifestations of simple but highly abstract underlying principles of grammar that differ only slightly across tongues. Work on languages related to English—such as Dutch, French, Spanish, and Italian—and nonrelated languages—such as Japanese, Chinese, Arabic, Hausa (of West Africa), and Warlpiri (of Australia)—support this view. Data on a wide variety of languages, as well as observations about how languages are acquired, are expected to contribute in a major way to the development of a viable formal language learning theory. In particular, these data will permit the testing of various proposed theories of the learning process, including computer simulations of language acquisition, models evaluating the chances of learning a grammar from observing a modest sample of the language, efficient forms of artificial intelligence, and explicit behavioral models of language performance and proficiency. Grammatical and word acquisition studies in children should help to resolve controversies regarding the species-specificity of language, language-specific innate constraints, and the relative contribution of the child and the environment to the learning process. The improvement of communication between people and computers by developing programming modes closer to "natural language" will also benefit from this work.

A major research task in the next decade is to explore more completely the hypothesis of grammatical universality, through intensive investigation of languages that are just now being fully described and are historically unrelated to the commonly studied ones. This work will involve considerable coordination and a stronger international basis between theoretical linguists and those descriptive linguists whose research is directed toward these frontier languages. It is also vitally important to this work to add to the relatively sparse instances of longitudinal studies that cover the same people over a period of many years and across a variety of linguistic learning environments.

Machines That Talk and Listen

Shortly after World War II, attempts were initiated to devise computer schemes to translate from one language to another, to recognize spoken and handwritten language, and to convert text into natural-sounding speech. Abortive and highly expensive early approaches to automatic machine translation and speech recognition were made in electrical engineering projects, but they did not address the complexities of human language; participants in these efforts came ruefully to appraise their results as "language in—garbage out." After engaging linguistic and phonological expertise to diversify the lines of research undertaken, team efforts ultimately did make good progress, leading to the scientific and practical successes seen to date.

At present, interdisciplinary research aims to explain the intricate relations that hold between language, the world, and intelligent systems (whether natural or artificial). In one line of research, formal linguistic models are being developed that make explicit provision for the varying computational requirements of language understanding. Relating linguists' grammars to existing computer systems enables computer scientists to provide a new generation of interpreter programs that run much more efficiently than their predecessors. A new theory of the semantics of programming languages promises to unite two formerly separate analyses: the denotational meanings of terms in natural language and the functional meanings of instructions in a computer program. Other developments in this work include knowledge about the recursive (looping-back) nature of computational processes, symbolic systems, and the importance of shared knowledge and beliefs in conversational face-to-face interactions.

In recent decades, collaboration between linguists, communication engineers, and computer scientists has led to dramatic increases in knowledge and new methods for analyzing and synthesizing acoustic speech signals. The computer revolution has made it possible to acquire and analyze in hours or days, rather than months or years, the large phonetic data bases needed to study the sound structures of language. As a result of the investigation of intonational and other phonetic properties of speech, undertaken with the goal of constructing machines that talk and listen, there is a better understanding of how the units of a linguistic system relate to acoustic signals, leading to more natural-sounding, machine "voices" and more capable machine "ears." However, the present ability to synthesize speech far exceeds the ability to automate speech recognition and understanding, because the normal acoustic flow of speech cannot be readily divided into neat, uniform segments corresponding to discrete sounds, syllables, words, or even phrases, as is readily apparent by listening to unfamiliar languages. At any instant of speech production, several articulators (larynx, tongue, velum, lips, jaw) are executing a complex, interwoven, rhythmical pattern of movements, blurring the boundaries between words as well as between the phonemic segments that compose words. The problem is further complicated because each perceived unit—sound, syllable, word, or phrase—varies widely with phonetic context, stress or emphasis, and the rate of speech, style, dialect, and gender of individual speakers. (The additional problem of discriminating a speech signal from background noise, including other speech, was noted above.)

What is invariant that enables people to recognize and understand speech? The discoveries of the last several decades have contributed to speech-recognition schemes, but they are still far from satisfactory; since they can deal at most with 5,000 words, compared with the 14,000 known by a 6-year-old or the 100,000 to 150,000 used by monolingual adults. The current solutions are still not very subtle. It is expected that dramatic improvements will accompany increased understanding of how the brain carries out the segmentation and analysis provided by the linguistic system.

Reading

Reading incorporates a complex hierarchy of skills, each raising a long list of questions in its own right. What are the fundamental perceptual processes involved in registration of visual print? How does word recognition occur? In the absence of aural and gestural cues, how do readers accomplish phrase interpretation? How are processes of inferential reasoning, critical thinking, and behavioral response catalyzed by the text? How predictable are these responses?

There have been fairly dramatic advances recently in understanding each of these interactive processes linking reader and text, especially as a result of developments in the cognitive sciences. These advances have yielded new insights into what the eyes and brain do that enable a reader to comprehend written language. These insights have progressed to the point that computer simulations can pinpoint exactly where and why a given reader (with known skills) encounters difficulty in a text and disclose what might be done to avoid or ameliorate those difficulties.

Knowledge from this research has led to educational strategies that result in improved reading skills in children. For example, in some projects, children who are poor readers have been taught particular cognitive strategies discovered in the skilled adult reader, such as how to identify and retain the most central information in a text. Children given this training show large and general improvements in comprehension. Other studies are implementing new programs to teach word recognition, taking advantage in some instances of computer-based technology for practice, drill, and individual tailoring of the curriculum. The research has practical importance for scientific and technical documentation, display technologies, and instructional methods, and it also provides a major opportunity to expand the frontiers of knowledge about interactive multilevel learning.

Decoding and Dyslexias

The process of decoding letters and words is critical in learning to read and has received a great deal of attention over the years, constituting the bulk of the research relating to reading. One central issue in research and pedagogical practice is whether a reader must use the printed words to retrieve a sound and then use the sound to retrieve the meaning of the word. Most of the evidence shows that a mature reader needs an intervening phonological code only if a word is unfamiliar or if the material is particularly difficult. But phonological decoding is critical when first learning to read. For example, awareness of the phonological structure of language is one predictor of the rapidity of early reading progress; 5-year-old prereaders who can segment a spoken word into its constituent phonemes (who can, for example, follow an instruction to say "table" without the t) tend to be better at word recognition at the end of second grade. Such results reinforce the importance of a phonics approach at the earliest stages of reading instruction, when fluency of word recognition is the key factor to rapid progress.

By third or fourth grade, children have generally mastered recognition skills sufficiently so that individual differences in reading skills no longer closely reflect differential word-decoding abilities. At these stages, knowledge of particular kinds of text structure, such as narrative, becomes increasingly important to comprehension, and these matters are now receiving greatly increased attention. Even very young children are aware of narrative conventions and use this knowledge in understanding stories that are read or told to them. Recent analyses of grade-school materials suggest that many selections violate conventional story structures, making it difficult for a learning reader to establish the coherence of the story. Research on reading comprehension will lead to better selection of useful texts for teaching.

Some grammatical structures are known to be used much less in speaking than in writing. For example, cleft constructions like "It was John who won the trophy" are rarely spoken in English; rather, vocal stress on the name—"*John* won the trophy"—is sufficient to convey that the identity of the winner is the key information in the sentence. These differences between reading and speaking seem small, but for some dialects they may result in serious interference between the spoken language and comprehension of the written language. Such difficulties have been noted in some speakers of black English and some deaf children, for whom written English primers are virtually samples of a foreign dialect. While learning to read a second language without first knowing how to speak it is not uncommon for adults who are already literate in their native speech, it is a very unusual and difficult hurdle for children who are first learning how to read.

Explaining the wide variation among early and middle readers in the rates at which basic skills become automated and more advanced ones develop and finding the causes behind reading disabilities (dyslexias) are major challenges. Research on reading disability has proven to be especially difficult. There is still no general consensus about the nature or definition of dyslexia, and indeed there are probably several distinct kinds of dyslexia, some of which are corollaries or causes and some of which are effects of reading dysfunctions. The field has discarded numerous theories that have not stood up to close testing, such as the hypothesis that the disorder is visual, involving reversals between letters such as *b* and *d* or words such as "saw" and "was"; that the disorder involves particular difficulty in associating visual and verbal elements; or that poor readers have difficulty in maintaining information about sequences. Other theories that have some support, but not full assent, include the idea that there is a deficiency in the specifically auditory-linguistic background and that the problem results from contracted vocabulary, low verbal fluency, inappropriate grammar or syntax, or difficulty or slowness in word retrieval. The most promising approaches to these tangles of cause and effect appear to be continued fundamental research on normal acquisition of reading skills, more detailed examination of disabled individuals by interdisciplinary terms—psychologists, cognitive scientists, medical researchers, and educators—and more longitudinal research, starting with prereading children.

Interpretation and Comprehension

A number of techniques for studying human reading skills have in recent years achieved high levels of sophistication, enabling progress in areas that had

Typical College–Level Reader

(1,566) (267) (400) (83) (267)   (617)   (767) (450)   (450) (400)

Flywheels are one of the oldest mechanical devices known to man. Every

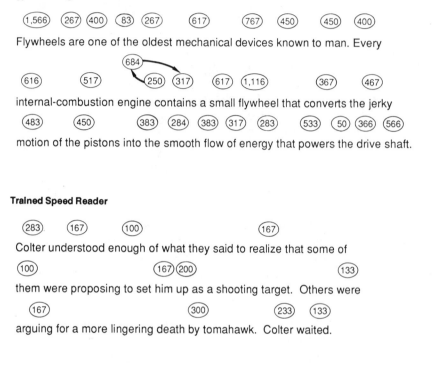

(616)   (517)   (684) (250) (317) (617) (1,116)   (367)   (467)

internal-combustion engine contains a small flywheel that converts the jerky

(483)   (450)   (383) (284) (383) (317) (283) (533) (50) (366) (566)

motion of the pistons into the smooth flow of energy that powers the drive shaft.

Trained Speed Reader

(283) (167) (100)   (167)

Colter understood enough of what they said to realize that some of

(100)   (167)(200)   (133)

them were proposing to set him up as a shooting target. Others were

(167)   (300) (233) (133)

arguing for a more lingering death by tomahawk. Colter waited.

Dyslexic Reader

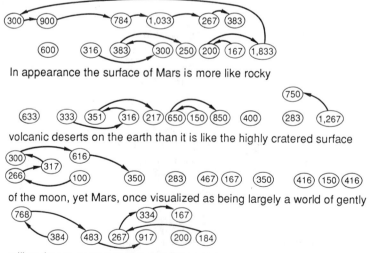

(300) (900) (784) (1,033) (267) (383)

(600) (316) (383) (300)(250)(200) (167) (1,833)

In appearance the surface of Mars is more like rocky

                           (750)

(633) (333) (351) (316) (217)(650)(150)(850) (400) (283) (1,267)

volcanic deserts on the earth than it is like the highly cratered surface

(300) (616)
(266) (317)
   (100) (350) (283) (467)(167) (350) (416)(150)(416)

of the moon, yet Mars, once visualized as being largely a world of gently

(768)   (334) (167)
   (384) (483) (267) (917) (200) (184)

rolling dunes, seems to possess little sand.

seemed quiescent for many decades. One such technological advance is eye fixation research, which provides a detailed characterization of where in a text and for how long a reader looks, pauses, skips, or looks back. This kind of research was impossibly cumbersome before the availability of laboratory computers and precision optical and video devices and is providing clear answers to questions that have been on the table since early in this century. Eye fixation

---

READING   How do a reader's eyes move across the written page? What does the mind do as the eye moves? How are marks on paper interpreted as words that convey ideas?

Researchers have developed methods to mechanically track a person's eye movements while reading a text and to meticulously analyze the patterns of gaze and subsequent accuracy of comprehension of the material. Such experiments are used to test and develop theoretical models of how people process what they see on paper. The different gaze patterns in this figure, represented by the pause durations (in milliseconds) and retracking lines, are typical of different kinds of readers. (There is no significance to the different texts shown; they are standard passages used in experiments.) Though texts differ in complexity, a given reader processes them the same way and at relatively small variation in overall speed.

Most people sample a written text densely, fixing their gaze (for more than 50 milliseconds) on about two-thirds of the words and rarely skimming past more than one word at a time; in most cases, the skimmed words are articles, prepositions, and conjunctions, "function words." Nearly all readers interpret each word immediately, inferring its meaning even when important clues to that meaning come later in the text. The duration of gaze on each word depends on the word's length, familiarity, and grammatical or contextual clarity. There is also a tendency to pause longer on the word at the end of a printed line—rather than at the end of a sentence—to "wrap up" any loose ends in interpretation. Normally, readers retrace their gaze only when a grammatical or semantic ambiguity leads them to rethink how to interpret a phrase.

Readers trained to "speed read" fixate on fewer words and spend less time on each fixation. Like normal readers, they retain very little information from words they do not fixate. The major skill that speed readers learn is to infer the meaning of a line of text based on a smaller sample of the words in it than before and not to tarry or retrack in order to resolve local ambiguities in meaning. The cost of these rapid inference procedures is less accurate comprehension, especially for nonrecurrent details.

Dyslexic readers seem to have trouble mainly in decoding written words, that is, segmenting the letters into syllables and retrieving their sounds correctly. These readers spend extra time on each word and frequently retrack in order to reinterpret inaccurately perceived words. And even at their slower reading rates, dyslexic readers generally develop a much less accurate understanding of a written text than other readers—a problem they do not necessarily have in interpreting spoken words.

research has revealed that the instantaneous perceptual window during reading is only two or three words wide even for highly literate readers, and that readers fixate on 70 to 85 percent of the content words of an unfamiliar or technical text, but only 40 percent of the short function words, such as "an" and "of." Also, readers attempt to understand the meaning of each word as soon as they see it rather than postpone a choice of interpretation until more data (from the remaining parts of the sentence) are collected. If the interpretation proves incorrect, readers then seek the source of the problem and choose an alternative interpretation that works better. The characteristics of each content word, such as its familiarity and the moment-to-moment demands of linguistic and task processing, control the duration of pausing on the word. Pause duration is thus a gauge of processing load or difficulty at each point in a text.

Since a reader is constructing a mental representation of a text while processing only a single phrase or sentence at a time, short-term memory limitations may play an important role in reading comprehension. This is not a new hypothesis. But the original theoretical claims that short-term memory may be an important bottleneck for comprehension processes were contradicted by the fact that standard short-term memory tests (for example, digit span) did not correlate with reading comprehension skills. The more recent experimental work does not evaluate short-term memory capacity in static contexts or at nonreading tasks; rather, it evaluates how much excess capacity different readers have available during the actual reading process: the resulting measures are highly predictive of comprehension levels.

Detailed temporal analyses of this sort have also made it possible to write a computer program that reads newly encountered though very carefully edited and prepared texts about as well and as rapidly as human readers. The program "pauses"—taking longer to process a passage—when there is difficulty in comprehension and speeds up when comprehension is easy. It can produce a summary or answer certain kinds of questions about the content of texts about as well as humans can.

Efficient readers adapt their expenditure of effort and ingenuity to their goal in reading, such as "light" reading for relaxation, reading to learn, or reading to react to an action document, such as a request for funds. Mature readers possess a complex repertoire of reading and study strategies for enhancing understanding and for detecting and overcoming comprehension difficulties. Considerable advances have been made in understanding of how these learning activities develop and how they can be enhanced by instruction. All of these advances can now be described and measured in terms of scientific theory rather than just intuitively, as was once the case.

No artificial intelligence reading program comes close to the seemingly effortless way in which humans seem able to call up just the right information needed to interpret a text and ignore the mass of irrelevant data. But recent work has begun to create substantially better insight into how people do this, and one interesting fact is that readers do not simply ignore irrelevant information. Consider the sentence: "Thieves broke into the vault of a bank and stole a million dollars." Do you even consider interpreting "bank" in the sense of "river bank?" No? Sophisticated methods show otherwise. If a person is flashed the word "bank" and then, immediately, "money," the response to "money" is about 40 milliseconds faster than when an unrelated word (say, "bark") comes first. This is called a priming effect. In a context like the sentence above, if "bank" is followed quickly by "river," instead of "money," the priming effect on recognition is present for this word as well, showing that both possible meanings had been activated. After a 300 millisecond interval, however, priming would occur only for the contextually appropriate meaning ("money"). The results of such laboratory experiments suggest that a rich knowledge of relationships among words may facilitate decoding and perhaps other levels of processing, even though there is no conscious awareness by a reader of how such knowledge is being processed in the course of reading.

A recent goal of language research is to develop a scientific basis for writing comprehensible texts. Previously, the classical view was that texts are less readable when they contain long sentences and use uncommon words. But research shows that revising a difficult text by shortening its sentences and simplifying the vocabulary does not make it substantially (or sometimes even at all) more comprehensible. One idea being studied is that some texts may be hard to read because the reader must repeatedly search in long-term memory for specific information needed to interpret phrases or sentences. Other research focuses on how well the syntactic devices and cues in a text focus the reader's attention on its major themes or most important information. Related research has shown that successful didactic writing highlights information that the reader needs in order to act or that sets forth specific examples that the reader will most likely have encountered or expect to encounter.

Continued work along these lines should lead to improved and simplified models of how to make written documents more comprehensible, but this is only a proximate goal. Making a text comprehensible for a given reader will certainly contribute to understanding, but it will not necessarily lead to learning. People can comprehend, remember, and summarize a text and still be unable to use the information acquired; learning requires the integration of textual information with previous knowledge. There is much work to be done to discover how people learn, how they use the information acquired from a text in new situations.

OPPORTUNITIES AND NEEDS

The results of research on basic processes linking the mind, the brain, and behavior have grown impressively in the recent past and show even more promise in the immediate future. Some of those results have led or soon will lead to valuable applications, such as new types of photography; economical telephone-line transmission of video pictures; better hearing aids; ways to improve normal and impaired memory; enhanced computer abilities to synthesize, decode, and translate natural speech; vastly improved expert systems to aid in such matters as medical diagnosis and car repair; and innovations in teaching students across all areas and levels of learning. But the major contribution of this research is that of sheer knowledge about human beings and other intelligent beings—a wealth of insights into the nature, possibilities, and limits of intelligent individual action. In this section we propose increased expenditures of approximately $61 million annually for this research: for equipment, investigator-initiated grants, new data collection, research centers, predoctoral and postdoctoral fellowships, postdoctoral training institutes, and multidisciplinary collaborative activities.

Research in these areas depends heavily on instrumentation for simulating, modeling, and controlling experiments, generating stimuli, recording and analyzing data, and developing new theoretical models, which require substantial amounts of computational power. Some of this equipment is currently only available, if at all, in such special settings as clinics, national laboratories, or expensive commercial centers. Increasingly, there is a call for greater access to powerful workstations and supercomputers. There is also a lively expectation that massively parallel computer architectures will be especially well suited to many behavioral and cognitive research problems. To some extent the computational need is filled by machines available at major research universities, but during the recent period of serious funding cutbacks and stringencies imposed on the behavioral and social sciences and the increased regulatory demands on research involving animals and humans, the overall level of instrumentation in laboratories has fallen seriously short of the research needs. Many formerly up-to-date university laboratories are no longer adequately equipped to do the research that is now possible.

In light of the increasingly advanced and specialized technology required to carry out experiments on behavior, mind, and brain, an estimated $25 million in new funds annually are needed to acquire or provide access to new equipment. We estimate that about one-half of this amount ($12 million) should be allocated to new and upgraded laboratory equipment and service facilities, about one-fourth ($7 million) to new computer hardware and software development, about $4 million to the improvement of animal care for research animals, and about $2 million for access to major neuroimaging devices.

The principal mode of studies on behavior, mind, and brain is carried out by small groups of investigators: one or two principal investigators working on a separately funded project (though each investigator may have more than one project) with one or a few assistants, who may range in level of training, skills, and independence from technicians to postdoctoral scientists. This approach has worked well, and we do not recommend major changes.

Such research by small groups is largely supported by investigator-initiated grants. The aggregated funding level of these grants has been somewhat buffered from roller-coaster trends in behavioral and social sciences spending at the federal level, due partly to the linkages and overlaps with life science and computer science research, which have been on more monotonic funding paths. Cognitive and behavioral sciences support nevertheless has not kept up with the growth in scientific opportunities, such as the rapid shifts in technological capabilities that have resulted from the microprocessor revolution. A rapid increase in funds for investigator-initiated grants—by about $20 million annually—is a high priority. These funds should be used to increase research productivity in three ways. First, they should be used to reverse the lack of change or actual reduction in the size of grants that has occurred at a time when real costs are increasing. Those real cost increases have resulted from improvements in animal care, the need for auxiliary staff in experiments involving infants and children, and the routine costs of supplies and services, among other matters. Second, they should be used to increase the duration of a typical grant to an experienced investigator from 3 to 5 years. Such an increase will sharply reduce the burden of time imposed both on investigators preparing renewal grant proposals and on the individuals asked to read and evaluate each proposal (generally 6, but sometimes as many as 15), which is a nonnegligible cost in research time. Third, they should be used to increase somewhat the total number of grants available because many promising proposals are now being turned down due to lack of funds.

A great deal of the pioneering research lies at the interface of disciplines and often requires a great deal of highly specialized technical expertise in fields ranging from biochemistry and physics to neuroscience, physiology, psychology, linguistics, economics, statistics, mathematics, and computer science. This inherent feature of much current work on behavior, the mind, and the brain calls for a number of developments to foster and facilitate more collaboration among people from very different backgrounds. This facilitation may take different forms, such as short-term visits, new centers of research, interdisciplinary training in the predoctoral curricula of universities, or the growth of postdoctoral programs that broaden rather than intensify the focus of new scientists' thesis work. We especially note the need for advanced training institutes for younger postdoctoral-level researchers working on newly emerging methods in highly technical specialities and recommend an additional $1 million be spent annually for such institutes. Given the geographical dispersion of investigators, collaboration can often be sustained only if there are opportunities to come together periodically in joint working sessions. Short-term workshops, seminars, and conferences focused on exchanges of current ideas and methods are the ideal medium to encourage this, and we recommend that $1 million be added to current levels of funding for such activities.

The picture of recruitment into graduate schools in the behavioral, cognitive, and brain science disciplines continues to reflect a growth trend. But much of this growth leads toward clinical and other nonresearch employment, and there is growing concern about the number of highly motivated and talented young people who are entering research careers. In particular, there is a relative lack of opportunity for individuals to gain intense experience at the predoctoral or postdoctoral levels in a variety of research methods and theoretical disciplines. This is only in small part a problem of curriculum: it is in large part a problem of funding, which is available more for teaching and clinical or applied work than for research apprenticeships. We therefore recommend that an additional $5 million annually be invested in research fellowships at the predoctoral and postdoctoral levels, with the majority share ($3 million) directed to the post-doctoral level.

There is in certain areas a critical need to inaugurate new data bases that can be accessed by large numbers of investigators; in particular, longitudinal data bases concerning cognitive and educational development. The cost of these new data collection effort would be about $5 million annually.

Finally, there is at present momentum in certain fields toward major new interdisciplinary research enterprises. In the recent climate of restrained budget growth, it has not been possible to organize the sort of stimuli—either in the form of planning grants or competitive announcements for new center programs—that would lead to competitive, full-fledged center proposals. We recommend that there be an initial commitment of $4 million to create two to three new interdisciplinary centers, with a view toward increasing this by as much as threefold if the experience of productivity warrants. We should note that some increment in support staffing in research agencies is a necessary element to generate, evaluate, and monitor good center grants.

2

Motivational and Social Contexts of Behavior

2

Motivational and Social Contexts of Behavior

On any list of plights of the human condition, one would find prejudice, addiction, violence, and suicide. They represent the extreme and morally most troublesome phenomena concerning the origins and regulation of appetites, purposes, and sentiments. Questions about these phenomena often lie close to the surface of everyday life, and they have been addressed since the earliest times at that level, frequently in texts of religious, ethical, legal, and philosophical reflection. The questions take such practical forms as: Why do strong emotions arise? Why does intoxication hold such overwhelming appeal to some people but not to others? How is criminality bred and how can it be contained?

Long-established traditions of reflection have yielded many compelling answers to such questions, forming the core of commonsense wisdom about them. But together these answers are often found to be contradictory. (Common sense, schooled by life, is greatly tolerant of logical contradiction.) Occasionally, one answer achieves predominance over its competitors, usually by acquiring the cloak of religious, ideological, or political authority.

Research in these areas thus has peculiar burdens and special responsibilities. One burden derives from the power of common sense to insulate us from surprise. Many scientific theories about familiar matters, when proven by empirical research, are deemed obvious, fully expected, and not having needed scientific verification. Yet common sense could also have deemed directly contrary findings to be obvious and fully expected. The special responsibility of research is to remain skeptical, to be unwilling to accept too readily any result that accords with common sense, to insist on scientific proof. A more significant burden is that research findings may not support a currently dominant idea. The responsibility of research is to follow theoretical leads and empirical results wherever they lead.

In this chapter we discuss research on affective states and processes; the linkages between health, behavior, and social contexts; the causes and control of violent crime; and the nature of social interaction. Among the matters now under intensive study are the social and motivational conditions that affect vulnerability to depression, cardiovascular disease, and risk of addictive behavior; the competing environmental cues and psychobiological processes that affect eating behavior and body weight; the ways that parental practices in managing children and criminal justice practices in sentencing offenders affect the course of "criminal careers," and the manner in which the social origins, size, and divisions of responsibility in a task group can affect what it produces. The research methods involved in these areas of investigation range from biophysics to cultural analysis. Research on motivational and social contexts of behavior, perhaps more than any other area in this report, accents the promising opportunities and the sharp challenges of multidisciplinary science.

AFFECT AND MOTIVATION

How and why do drives and desires grow and decline? Are certain human emotions primary and universal? By what processes do children reach emotional maturity, and why do some fail to do so? How do people and animals regulate their appetites under changing conditions of scarcity or abundance? Why does exposure to stable environmental circumstances not stimulate identical or consistent behavior?

These and similar questions animate research on motivational systems and affective states and processes. At the physiological level, there is research, primarily using animals, on the motivations associated with hunger, thirst, and sexual behavior. By studying animal behavior in conjunction with brain events and structures, significant progress has been made in uncovering the neural features responsible for these motivational systems. There have also been major advances in understanding homeostatic systems that operate to control the intake and expenditure of energy and, in humans, the neuromuscular codes that make the face a prime vehicle for emotional expression.

At least as important as these advances at the physiological level is better recognition of the complexity of motives and emotion. These phenomena cannot be reduced to the workings of single variables. The relations between cognition and emotion are complex, and emotion can be understood only in context, which almost always involves the cultural meaning system in which behavior is embedded. Moving from primarily individual phenomena, such as pain and hunger, to more interactive phenomena, such as indignation and envy, the cultural components become more and more compelling. The understanding of affect and motivation thus draws on talents and skills ranging from those of chemists, physicists, and computer experts staffing brain-imaging facilities, through biomedical and behavioral specialties traditionally concerned with personality and mental illness, to sociologists and anthropologists studying the contexts, occasions, and meanings of emotional expression.

Emotional Expression, Perception, and Maturation

Identification of the facial, vocal, and postural correlates of different emotions and development of increasingly sensitive measures of the activity of peripheral and central nervous systems hold important promise for illuminating the course of normal emotional development. They also provide major new opportunities to investigate the natural history, diagnosis, treatment, and prevention of affective illnesses, such as depression.

One new approach that has already proven useful involves measuring the relative tension of various facial muscles that affect the facial expression of emotions. This muscle tension can be monitored electronically by surface electrodes, and the corresponding facial expressions can be monitored by television cameras and evaluated by anatomically based coding systems. Techniques of electronic recording, display, and analysis of the action of all the muscle groups that contribute to expression in the face now enable precise quantitative descriptions and realistic computer simulations of emotional expression. These and other advances have permitted detailed study of the development, perception, and decoding of emotional displays and explorations of deceptive communication of emotion. Such information can provide, among other applications, a new empirical basis for testing and monitoring the utility of pharmacological and other strategies used in the treatment of affective illnesses.

Recent research indicates that a small number (5–10) of facial expressions seem to have the same meaning cross-culturally. The role of such communications in regulating social order, as well as more elaborate learned expressions, is now under study. Communications of affect—"nonverbal communications"—are an important part of socialization since this is a powerful way for peers and elders to convey their attitudes about objects and behaviors of importance in their culture. Researchers are beginning to use technical advances in measuring facial expression to aid in studying the development of children's competence at deciphering expressions and to see how this form of communication regulates behavior and stimulates emotional growth.

For example, a new "social referencing" paradigm is being used to study infant social-emotional sensitivity. An experimental demonstration of this paradigm in uncertain or potentially fearful situations involves the response of 1-year-olds to a "visual cliff," an illusion of discontinuity created by shifting grid

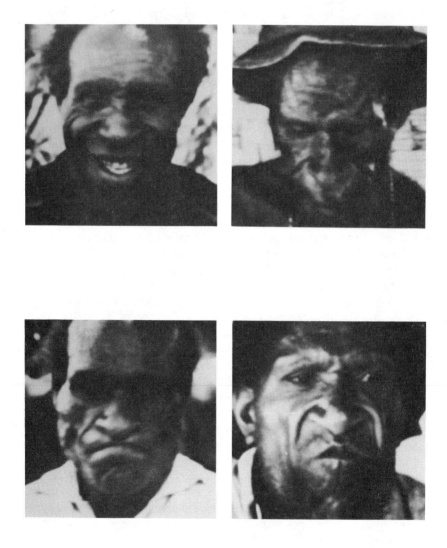

patterns laid over a flat surface. When the depth of the visual cliff is intermediate and the infants might or might not cross it, they look to their mothers. When mothers pose a "fear" expression, none of the infants cross; when mothers pose a "happy" expression, nearly all cross. In another experiment, when infants' mothers respond to a stranger in an unfriendly rather than neutral or friendly manner, the infants respond to the stranger with cardiac acceleration (characteristic of fear), less smiling, and more distress. This kind of evidence not only reflects infant sensitivity to emotional expression but also the use of that sensitivity in regulating behavior.

Recent studies indicate that young children up to 3 years distinguish primarily a positive and a negative emotional category with little differentiation, for example, among excited, happy, or proud, and among sad, angry, or afraid. Over the next several years children begin to differentiate within the positive and negative categories in ways that are in some respects universal and in others culture specific. One intriguing finding is that children only gradually—between 3 and 13 years of age—come to realize that it is possible to simultaneously experience different emotions, such as being happy that one's lost dog has come home but sad that the dog has been injured. At first, only the positive or negative emotion is acknowledged; later, children will typically say that the two emotions can be experienced at the same time. Research is proceeding on how learning of this sort in childhood may contribute to the emotional capacity to cope with the stresses, complexities, and responsibilities of life.

A developing mosaic of findings is beginning to yield a detailed picture of the link between a child's interaction with its parents, the child's emotional

---

FACIAL EXPRESSION   Are facial expressions of emotion entirely culture-specific, or are there some universal expressions across the human family? If there are universal expressions, what are the common muscular gestures of the face that comprise them and what emotions do they express?

Researchers interested in these questions have traveled far, in this case to mountain hamlets in New Guinea whose inhabitants have had minimal contact with the outside world—the people have never worked for Westerners or traveled to commercial centers; neither speak nor understand English or pidgin; and have never seen photographs, magazines, or motion pictures that might convey outsiders' modes of expression.

In the video frames shown in this figure, the instructions were to pose emotions for: "your friend has come and you are happy"; "your child has died"; "you are angry and about to fight"; and "you see a dead pig that has been lying there for a long time." The facial displays corresponding to these emotions—happiness, sadness, anger, disgust—are easily recognized by and the same as those of Westerners. A variety of experiments similar to this one have been done in other cultures. In fact, the four emotions shown here and a few others (fear, surprise) are easily and consistently produced and recognized across preliterate and literate cultures.

perceptiveness, social competence, and peer group status. For example, a number of researchers have documented the principal way that fathers and mothers differ in their style of child play, mothers being more verbal and didactic while fathers are more physical and arousing. In turn, these stylistic differences are reflected in children's social development, with the children of highly verbal mothers and physically very playful fathers being rated more socially competent with their peers. Other recent work indicates links between a child's ability to accurately recognize emotional expressions and the child's social status among peers; children who more correctly identify facial expressions are more popular.

Some evidence further suggests that early peer status is predictive of later social and emotional adjustment. For example, there are findings that different types of low peer status vary in their stability and degree of association with later adjustment problems; specifically, the status of rejected children is more stable over time than that of either neglected (by peers) or popular children. Another line of work indicates that unpopular children do not necessarily differ in their knowledge about what to do in various social situations, but they do differ in the way they construct goals: popular children have more friendly, assertive goals, while unpopular ones have more task-oriented goals. Moreover, children who attribute social rejection to their own personal inadequacies are less likely to cope effectively by changing their styles than those who do not. Pilot intervention studies focusing on these and related dimensions of social competence are proving effective in promoting change in the social status of some socially rejected children. There is obvious promise as well as challenge in trying to map out the antecedents of emotional expression and discover how it operates on a fine-grained level to regulate social behavior and contribute to social competence.

Emotive Circuitry and Metabolism in the Brain

Researchers are beginning to gain very detailed knowledge about the nature of the different brain circuits involved in such motivational and affective processes as aggression, eating, drinking, and sexual behavior. The classical studies in the 1950s that proved that animals will work for electrical stimulation of certain brain areas to the exclusion of all else. They had an enormous impact. Animal studies descending from this pioneering work have shown that there are separate motivational systems for defensive fighting (between members of the same species), predatory fighting (usually between members of larger and smaller species), and feeding, but that the neural mechanisms of predatory attack are more closely related to the mechanisms of feeding than to those of defensive fighting.

Motivational systems passing through the hypothalamus appear to be activated at different stages by food and water, as well as by opiates and stimulants. The latter substances appear to activate the systems more strongly than the more biologically useful rewards; this finding is important for understanding biological support for addictive behavior. In other studies, researchers have determined that, for most right-handed people, the right cerebral hemisphere is more specialized for both expression and reception of emotions, in comparison with the specialization of the left hemisphere for language. The exact significance of this asymmetry is still unknown; it is as yet an intriguing clue as to how the brain organizes emotion.

Over the last few decades, a new discipline of neurochemistry has emerged. A wide variety of neurotransmitters, chemicals that transmit impulses from one nerve cell to another, have been identified—more than 50, with estimates as high as 250. Different chemicals seem to operate in brain systems to carry out different functions. For example, one category of neurotransmitters, the endorphins, are especially involved in the modulation of pain; another, the catecholamines, operate in the control of activation and mood and play a critical role in episodes of depression.

A variety of innovative technologies are making the anatomical and functional features of the brain more visible and quantifiable. Using new imaging methods, researchers can now measure neuroanatomical features, conduct studies of the volume and density of specific brain structures, and study the metabolism of living brains with unprecedented margins of safety and precision. Such projects involve interdisciplinary collaboration on a broad front. For example, high-quality PET studies of brain structure use precise cyclotron targetry, mathematically advanced techniques of data reconstruction, high-purity radiopharmaceuticals, advanced diagnostic techniques, ingenious neuropharmacological strategies, and sophisticated behavioral methods and assessments. These methods make possible the discovery and analysis of detailed brain circuits that underlie normal and abnormal affective processes, the localization of receptors that are affected by psychoactive drugs, and an ever-expanding horizon of related studies in humans and animals.

Biobehavioral Rhythms

Behavior that is rhythmic ranges from the relatively exotic, like the mating of marine organisms that takes place for only one day, at precisely the the same time every year, to the familiar sleep-awake cycle of most mammals. Rhythmicity is a process that affects almost every realm of psychological function, from diurnal changes in thresholds for discriminating simple auditory and visual signals, to monthly changes in human mood (correlated partially with the female menstrual cycle), to annual variation in breeding and hibernation.

Behavioral research on biological rhythmicity is advancing along two main fronts. First, there is a rapidly developing theory of biological cyclicity, especially daily (circadian) rhythms. This is a quantitative, formal set of models that holds for a wide range of organisms, from insects through mammals, including humans. Based on mathematical oscillator theory, this model has proven useful to understanding behavioral patterns as seemingly disparate as activity rhythms in laboratory mammals and cyclic seasonal depression in humans. A second front is the analysis of specific psychologically relevant behavior. Disturbances in sleep, depression, and other forms of psychopathology all have cyclic components. Perhaps more important, some syndromes may be caused directly by jarring alterations in normal cyclicity, such as can arise from long-distance air travel.

Patterns of Food Consumption

How do animals and people regulate their weight? If adult animals are experimentally starved or fattened and then given free access to food, they generally return to their original weight. One explanation for this result is that the brain senses the body's weight (perhaps monitoring the size of fat stores), just as it senses blood pressure or carbon dioxide, and is set biologically to "defend" a certain weight through shifts in food consumption or metabolism, similar to the way a thermostat is set to maintain a particular temperature. This set-point theory of weight regulation has been a fruitful point of departure for research. The theory suggests that most overweight people may happen to have a high set-point, so that maintaining lower weight is a constant battle between set-point regulation and psychological or social pressures to be thinner, leading to the bioregulatory pathologies of anorexia and bulimia and the frustrations of tens of millions of dieters.

But researchers are also focusing on the ways in which this view of eating behavior is incomplete. One line of work with animals and people indicates that eating often occurs not in response to a present weight deficit but in anticipation of one, implying an important role for learning. In the last decade, there has been a remarkable coming together of three disparate traditions of studying how food-seeking behavior adjusts to environmental opportunities: operant conditioning (learning for rewards), the economic theory of resource utility maximization, and functional approaches in ethology. In all three domains, the immediate issue is how animals and humans settle on more or less optimally rewarding behavior.

In operant conditioning, the issue is expressed in terms of how a pigeon allocates key pecks or a rat pushes on a number of different keys that distribute reinforcements (usually, access to food) according to different schedules (that is, temporal patterns of an irregular, chancy nature). In economics, the issue is generalized to how inputs pegged at various prices combine to yield the most satisfactory overall return or value. In ethology, the issue appears as the problem of understanding patterns of foraging in the wild for a variety of possible food sources.

In an operant experiment, an animal is left for some time in an environment where it can engage in a number of its usual activities, such as grooming and drinking, but in order to eat it must work by pushing on certain keys. The experimenter differentially rewards some patterns of behavior, and that intermittent schedule of rewards is found to be highly controlling of what the animal chooses to do with its time. The relationships between schedule changes and behavior changes can be expressed as mathematical regularities. When combined with information about an animal's nutritional needs and characteristics of food availability in its natural environment, these regularities have been shown under certain conditions to yield accurate predictions about foraging behavior. Ethology and operant conditioning, two research fields that were diametrically opposed in the past, have thus begun sharing ideas and methods. The laboratory operant box has become a tool of research on natural adaptation.

The laboratory situations studied by animal behaviorists seem to be, in some ways, more complex and more natural (in terms of simulating the conditions of free-ranging animal behavior) than the situations usually arranged for human research subjects. However, when roughly comparable naturalistic situations of activity allocation are posed to people, their behavior turns out to be well described by the same mathematical generalizations. Investigators of these different topics have recognized their common analytical interests and are now working together to develop predictions of behavior based on various possible fundamental principles. One very important issue is the empirical assessment of the economic principle of cost-benefit maximization in comparison with an alternative principle that derives from the animal studies and some of the human research: that attention is accorded to respective activities so that each yields the same average rate of reinforcement rather than the same marginal rate.

Eating must also be studied in the context of other behaviors; rather than experiments in which caged animals are given access to a food cup but little else to do, the present challenge is to design theories and experiments combining food availability with numerous other available activities (sexual behavior, caring for the young, other forms of interaction, and so on) as competing alternatives.

The study of foraging illustrates an important trend in the attempt to understand motivation: the merging of functional-adaptive approaches with the study of mechanisms. Claims about the adaptive value of certain behaviors— for example, the greater selectivity in choosing mates by females because of their greater investment of the offspring—have prompted research to test such sociobiological claims and to search for the mechanisms that mediate the behavior.

The senses of smell and taste are also important in regulation of eating. Particularly for elderly people, loss of appetite can often be explained by a failure of the sense of smell. And since the smell system is much affected by Alzheimer's disease, subtle alterations in smell sensitivity may be key elements in developing early diagnostic techniques. Research is also directed toward understanding genetically determined individual differences in taste. For example, two-thirds of people tested taste the compound PTC (phenylthicarbanide) as distinctly bitter; the other third do not. These two groups respond very differently to sweeteners, one finding them much more potent than the other. It should be noted that scientific and clinical interests in developing a comprehensive understanding of taste and smell are joined by practical interests in improving the prediction of sensory effects of food ingredients based on their chemical composition.

On a different track, researchers are working to understand the functions that eating serves aside from being a source of nutrition, most notably as a vehicle for social symbolism and communality. For example, in most of south Asia—which accounts for close to 20 percent of the world's population—food is a major medium of communication that reinforces the hierarchial nature of the social system. Rules about who eats what food handled by whom play a powerful role in supporting and defining the caste structure. To fully understand food selection, the many and complex functions of food need to be taken account of in behavioral theories.

BEHAVIOR AND HEALTH

Although it is still common to think about health as though it were strictly a matter of human biology—the research province of medical scientists, physiologists, pharmaceutical chemists—such thinking has become outdated. Until about 1940, the leading causes of death and disability in the United States were acute infectious diseases, such as pneumonia and tuberculosis; today, such diseases are not usually fatal, and the leading causes of death are illnesses that are significantly affected by behavioral and social factors, such as cigarette smoking, excessive drinking, illicit drug use, bad dietary habits, violence, stress, and refusal or inability to maintain recommended medical regimens.

Of the 2 million deaths in the United States every year, about 1.5 million are due to heart disease, cancer, stroke, or fatal injuries and overdoses; for all of them, behavioral and social causes are known to be major risk factors. Studies of life expectancy and mortality among cigarette smokers, former smokers, and nonsmokers yield estimates of more than 300,000 excess deaths a year due to smoking. Epidemiological studies in metropolitan areas indicate that one of every five adults suffers from a diagnosable form of motivational or emotional disorder during a given year, most commonly alcoholism or depression.

Fundamental research concerning the behavioral and social contexts of health and illness has increasing applications to prevention and treatment. Research taking place in laboratory experiments, accompanied by rigorous theoretical modeling, and in field sites such as schools, workplaces, and clinics addresses such questions as: How do social influences and motivational dispositions affect the onset, maintenance, and cessation of drug, alcohol, and tobacco use? How pervasively do environmental stressors affect the body? What are the social and behavioral factors and processes that increase or decrease vulnerability to disease? How do variations in the incentives and costs of health care affect how and whether people seek treatment, provision of care, and the outcomes of treatment?

Prevention of Drug, Alcohol, and Tobacco Abuse

"Peer pressure" and "youth culture" are bywords today of influences on adolescent alcohol, drug, and tobacco using behavior. This was not the case 20 years ago, when these ideas, borrowed from social science, first began to revise the popular and scientific thinking that linked youthful smoking to the straightforward desire to emulate adults, teenage drinking to juvenile rebellion, and illicit drug initiation to predatory pushers. Since then, a phalanx of careful studies, including ethnographic investigations among drug users in and out of school, panels of adolescents followed at yearly intervals through young adulthood, and repeated national surveys of successive classes of students, have revealed a far more detailed and accurate picture, suggesting bases for selecting among and developing new approaches to treatment and, especially, prevention of problems induced or aggravated by alcohol, tobacco, or drug abuse.

Contrary to earlier impressions that pushers cut some innocent children out of the pack and introduce them into a lonely world of drugs, drug use among school-age children has proven to be an intensely social activity. Researchers studying friendship patterns across time among secondary-school students in New York, for example, have found that the most closely correlated characteristic of close friends, aside from their age and sex, is their level of marijuana use. About half of the time, this pattern results from acquiring new friends with similar established patterns of drug use or nonuse; the other half of the time, existing friendships lead to subsequent matching of patterns of consumption or abstinence.

Experimentation with different substances seems to be related mainly to environmental opportunities and broad social factors. Student drug use in junior and senior high school is mostly a friendly or party activity, which varies depending on the demographic characteristics of the students in the school and the overall peer perception of the harmfulness of use. But the evidence impressively supports one long-suspected factor: young people reared with poor self-images are especially attracted to intoxicants, finding in them a common bond with other troubled youngsters, which may lead them together into increasing alienation from noninvolved peers and activities.

Young people's use of intoxicating substances is known to follow a remarkably standard sequence, going from alcohol to cigarettes, to marijuana, to pills or powders, to intravenous drugs. But moving any distance in this sequence is neither inevitable nor typical. In recent years, more than 90 percent of high school seniors have tried alcohol, 70 percent cigarettes, 55 percent marijuana, 17 percent cocaine, and 1 percent heroin. The proportions who use any of these daily is much lower: about 20 percent for cigarettes, 5 percent for alcohol, 5 percent for marijuana, and less than 1 percent for other substances.

A strong curvilinear association has been discovered between age and the prevalence of narcotic (opiate) abuse, with use beginning between the 6th and 12th grades, cohort use peaking by age 25, and among the continuing steady, heavy users, a rapid falling off after age 40. Adolescent friendship networks and general societal prevalence influence the onset of use. Formation of specialized supply-focused groups is important to its continuance. Later associative transitions, especially marriage, tilt strongly toward the termination or sharp reduction of use in early adulthood. In the older years, a combination of declining health, more serious judicial jeopardy, and the prospect of permanent loss of social supports, such as employment, careers, and family relations, where these exist, lead to fewer users.

Evaluating the Effects of Interventions

Research on the effectiveness of preventive intervention against early alcohol, tobacco, and drug use has advanced considerably in the past decade, yielding more detailed understandings of developmental sequences and their conditioning at crucial transitions in the contemporary life course, while offering possible ways to reduce the problem. Net positive results have been demonstrated in some research-based health education experiments that capitalize on the willingness of students to behave with appropriate self-regard in the presence of supportive peers. Such programs avoid the zealous overkill of many past programs and teach how to reject offers to share drugs, including alcohol and cigarettes, in socially graceful ways. These lessons have proven very credible when a distinction is made between the risks of frequent heavy use and those posed by trial or occasional use, especially when this information is received from respected and properly trained adults and older youths. Effectiveness is clearly increased when these lessons are reinforced through a variety of channels, including mass media and voluntary group efforts as well as school curricula. Some of these experiments have now been expanded and modified into major national campaigns and commercially published packages. However, the expansion and modification of these intervention techniques preceded expansions of evaluation research to determine if and how well the positive results hold up across different social environments and with other "teachers" than the original research-based intervention teams.

There has been a substantial gain in understanding how education and media can affect the use of cigarettes. An important statistical finding is that cigarette sales per capita decreased most during the period (1967–1970) when nationwide "equal time" anti-cigarette-smoking television advertisements were in place. This decrease coincided with several other factors: rising cigarette excise taxes, more general information concerning the medical risks inherent in regular cigarette smoking, and a shift in the normative acceptance of smoking and of the appropriateness of social intervention by physicians, family, and even strangers, based on increasing concern about health and fitness in general. Subsequent studies conducted on smaller geographic scales in rigorous, experimentally controlled programs—with special news features and direct educational efforts in small groups—have confirmed the positive effects of "health promotion" advertising aimed at preventing smoking among adolescents as well as adults. The key questions now concern the transferability of such efforts, the persistence in their efforts, and the design of more efficient ways to achieve these objectives. Further experimental and longitudinal research is needed to answer these questions.

Addiction and the Marketplace

The character of addictive substances as market commodities is a relatively new approach that has proved very useful. Recent studies of alcohol, tobacco, and even heroin use have clearly shown that laws of the market operate for each of these substances: when prices rise there is an overall decrease in the amount of substance consumed, although the total amount of money spent on the substance increases. Data are available from the 1960s and 1970s, when real alcohol and cigarette prices shifted from year to year because of the seesaw effects of higher excise taxes and overall price inflation that cheapened these taxes. These data suggest a substantial negative relationship between price and consumption (elasticity figures of about -0.5, that is, a decrease of one-half percent in consumption for each one percent increase in unit price) for alcoholic beverages and cigarettes. These elasticity results have important implications for public health policy, because the level of consumption is closely related to rates of illness and death due to lung cancer, cirrhosis of the liver, drunk driving crashes, overdoses, and the like.

New data show how different age and sex groups respond differently to price increases (such as occur when excise taxes are raised). For cigarettes, the effect of price increases appears to be strongest among 15- to 24-year-olds, and it is stronger among women than men. In general, the effect seems to be through changes in "participation" (that is in smoking or not smoking) rather than in "amount" (that is, the number of cigarettes smoked by those who do). The effect of alcohol price changes, however, appears strongest among males aged 20 to 50 and involves amount as well as participation. Confirmation of these findings and more details about how price changes affect participation and amount, and how different social groups respond, could be very useful in forming public policies.

Prevention of AIDS

It is likely that between 1 million and 1.5 million Americans have already been infected with the AIDS (acquired immune deficiency syndrome) virus. At present in the United States, AIDS is concentrated among homosexual and bisexual males and users of illicit intravenous drugs, and there is also relatively high incidence of infection in children born to infected mothers, recipients of blood transfusions, and sexual partners of those in these high-risk groups. The epidemic is spreading much faster in the black and Hispanic population than in the white population. However, present evidence indicates a likelihood that the virus is capable of spreading to a very large proportion of sexually active individuals, well beyond the groups now identified as high risk.

The consensus among biomedical experts is that fully effective treatments and vaccines may not be available for at least 5 years and maybe considerably longer. In the interim, practical efforts to cope with the immense social, psychological, and economic problems of the AIDS epidemic will require contributions from the social and behavioral sciences in a number of areas, including ones that have been relatively neglected in recent years.

One of those neglected areas is basic research on human sexual behavior. There is a renewed appreciation of the need for vigorous programs of research on such behavior, which has always been a sensitive undertaking in American society and which has lacked extensive public investment. But understanding the variety of sexual behaviors and how people integrate those behaviors into their lives will be crucially important in persuading individuals to modify their sexual behavior in order to diminish the risk of spreading AIDS. Lack of knowledge about sexual behavior also seriously handicaps current efforts to predict the future course of the AIDS epidemic in the United States. For example, there is a remarkable lack of reliable data on the rate and character of sexual contacts for the national population and for groups with elevated risks of infection. Even simple estimates of the size of the current population of homosexual men rely on data collected by Kinsey in the 1940s, which were considered suspect then due to the biased samples in Kinsey's surveys.

There is an immediate, crucial need for better knowledge of human sexual behavior because of the AIDS crisis, and such research will have practical applications in many other areas, including fertility and population studies, teenage pregnancy, and treatment of sexual dysfunction. Better understanding of human sexual behavior will also substantially contribute to the study of other sensitive and powerful motivational and social factors that affect personal and health behavior.

A second research area that is given added urgency by AIDS is the communication of health information. The goal of public education about AIDS is to increase public knowledge and understanding of the epidemic and to encourage people to avoid behaviors that have high risks of spreading the infection. The most difficult challenges include producing effective and publicly acceptable education programs for youth; diffusing information to relatively hard-to-reach populations, such as cultural minorities intravenous drug users, and prostitutes; and providing information that fully informs without panicking the public, so that realistic public health policies can be debated and formed.

Massive and decentralized public education programs will have numerous unique elements and may involve a slow learning process with considerable trial and error. Rigorous evaluations of the effects of educational programs will be important in order to learn from experience and thereby improve the programs over time. The technology for conducting evaluation studies does not need to be created, but it does need to be implemented correctly. What is learned will not only be relevant to AIDS, but also to numerous kinds of educational programs and to knowledge about social influences on behavior more generally.

Three other lines of research acquire great practical importance due to the AIDS epidemic: (1) systematic historical study of earlier societal reactions to epidemic diseases, such as leprosy, tuberculosis, polio, and syphilis; (2) careful and continuously updated estimation of the costs and benefits of alternative intervention, treatment, and care strategies; and (3) microcultural studies of whether and how educational information, screening, intervention programs, and changing availability of devices, such as sterile needles or barrier contraceptives, may shift behavior to reduce the risk of transmission. A particular need under the last category is to determine how and under what contingencies individuals in different circumstances respond to learning the results of serum testing for AIDS infection.

Stress, Risk of Illness, and Behavior

How are changes in physiological functioning (for example, the endocrine and immune systems) affected by social and behavioral factors? Psychological stress, which has a variety of negative effects on health, may be mediated through the emotional system, and recent work suggests that stress can adversely influence the functioning of the immune system. Other research has found that social support mechanisms can, to a remarkable degree, reduce a person's sensitivity to certain stressors and increase compliance with medical regimens.

Psychological stress can markedly alter social roles and behavior patterns. For example, there is a direct association between bereavement and mortality in men, but not in women. Among the elderly, living alone is associated with poorer nutritional intake and poorer health for older men, but not for older women. Behavioral and social scientists have begun to struggle with the complex issue of how differing social support networks mitigate or contribute to a variety of disorders such as depression, schizophrenia, or cardiac disease. One line of research examines how the effects of social networks on health outcomes for the patient are mediated by psychological mechanisms such as the perception of personal control. This kind of factor appears far more significant than has been recognized by health care providers.

A rapidly developing area of research is the role played by risk, vulnerability, and protective factors in the etiology, maintenance, and recovery from such diverse mental conditions as schizophrenia, affective disorders, and (in children) attention-deficit disorder. Risk factors are those elements in the individual, the family, and the environment that are associated with higher incidence rates of such disorders in the population. Vulnerability refers to a condition of the individual, which is sometimes reflected by palpable, specific markers and sometimes conceptualized as a generalized or specific predisposition to pathological outcomes. The study of risk and vulnerability factors offers insights into the potential power of different predictive models of etiology of disease while providing, in turn, a rational basis for norms of preventive intervention with those individual who are prone to the development of mental and physical illness. Help-seeking channels and mutual-help groups are related contextual influences on health behavior that have come under increasing research scrutiny.

There are many studies on the deleterious effects of job loss, such as studies finding an increase in serum cholesterol in laid-off workers following plant closings. The most dramatic research traced the existence in some individuals of psychological depression for up to half a century after the economic depression of the 1930s and its job disruption. Most recently there has been increasing interest in studying the neuroendocrine correlates of stress, in an effort to understand the physiological mechanisms by which environmental events affect physical health. For example, recent biobehavioral research suggests that psychophysiological responsiveness (reactivity) to emotional stress may be a marker of processes involved in the development of cardiovascular disorder. Recent investigations have also examined sex differences in neuroendocrine responses to stress, as well as the effects of occupational stressors on adrenocortical and adrenomedullary hormones. A number of studies have begun to consider the regulatory effect of the central nervous system (CNS) on immunity. Under consideration are endocrine, neurotransmitter, and opioid influences as modulating processes, as well as a direct neuronal link between the CNS and immunocompetent tissues. In addition to regulatory mechanisms, attention has also been directed to feedback processes between the immune system and the CNS. These areas of research have been greatly facilitated by the availability of new bioassay and measurement techniques.

Behavior and Health Care Delivery Systems

Research on health care financing has taken on major importance as health care costs have risen to a level exceeding the national defense budget. A significant step in that research was a large-scale social experiment initiated in the 1970s, involving 7,700 individuals in six areas of the country, that yielded important information on the specific consequences of alternative forms of health care organization and services.

In one aspect of its design, the study compared the styles of medical practice (especially, hospitalization practices) of a health maintenance organization (HMO) versus a traditional fee-for-service system. Many observers view HMO medicine as cheaper, but skeptics have charged that this is an artifact of selection bias—namely, of enrolling generally healthier clientele from the outset. To test the issue, individual families were assigned randomly to each system, in both cases paid for entirely by third parties. Analysis of the results showed that the families in HMOs still had 40 percent fewer hospital admissions over a 5-year period than the fee-for-service group; this difference translated into a financial savings of 28 percent. The health-status consequences of this organizational difference and the relative satisfaction of the clients are now being analyzed.

On another issue, analysis of medical follow-up data obtained during the experiment showed that members of families who were assigned to health insurance plans that paid all of their medical costs made substantially increased use of services—about 40 percent greater—than those in plans with a substantial deductible (as high as $1,000 depending on family income). For the free-care group, there was measurable improvement in health outcomes for low-income people with certain problems, such as bad eyesight, decayed teeth, or, most seriously, high blood pressure. But for participants not afflicted with these particular problems, no overall improvement in health was evident despite the increased use of medical services. Much of the increased use, such as treating relatively minor injuries in the emergency room rather than at home (for example, cuts not needing stitches), was of minimal benefit. On balance, the data strongly suggested—no single experiment or study can be said to prove—that it would be cost-effective to improve health care by further subsidizing medical treatment for people afflicted with certain health deficiencies, but the complete removal of economic considerations in access to medical services would not be a cost-effective way to improve the health of the population.

CRIME AND VIOLENCE

The problem of crime has been in the forefront of national attention for several decades. Surveys have consistently shown that Americans rank it among the nation's most serious domestic problems. This has been especially true for certain serious crimes involving violence or invasion of the home—murder, forcible rape, aggravated assault, robbery, and burglary. Few demands on the behavioral and social sciences are more insistent than the call for knowledge on how to reduce the threat of these dangerous acts.

At the same time, criminal and violent behavior is a challenge and an opportunity for scientific research. Why do people help each other or hurt each other? Why do they voluntarily conform to some, but not other social norms? To what degree and under what conditions is behavior controlled by material or biological needs, character traits, internalized values, or calculations of moral, economic, or physical consequences? Competing theories of behavior face stern tests in having to account for the "deviant cases" of predatory crime in largely civil, law-abiding societies.

Because of the inherently stealthy nature of crime, direct systematic observation of it is rarely possible, and the systems for reporting crime have generally been weak and nonuniform. In recent years, however, the main source of data—the police-generated FBI Uniform Crime Reports—has been supplemented with two new sources: self-report surveys of individuals who, on a confidential basis, describe their own criminal activity; and victimization surveys, which report the experience of crime by the general population. Although all three data sources have limitations, they are now available for mutual validation and calibration, and it has been found that together they yield information that is consistent or can be readily reconciled.

Criminal Careers and the Effects of the Criminal Justice System

One of several promising approaches developed on the basis of interpreting these results in recent years has been to focus on criminal careers, that is, the development of criminal and noncriminal behavior as a life-long process. Although sporadic participation in crime is most common among adolescent males, nearly half of all urban-dwelling American males can expect to be arrested for some kind of nontraffic offense at some time during their lives. Yet all three data sources show that a relatively small number of "career criminals" or "violent predators" are responsible for a grossly disproportionate percentage of the crimes committed. A few people begin criminal activity quite young— some as early as 8 years of age. The highest rate of criminal activity in the general population occurs in the mid to late teens. But there is a heavy dropout rate from adolescent criminality, so that by age 30 or so those still committing violent crimes are largely the very serious, resistant, high-rate offenders who are unlikely to abandon their criminal activities much before age 40, if then. Research has not yet provided information about the dynamics and reasons for termination.

There is a strong nexus between drugs and crime. Regular users have much higher rates of criminal activity than nonusers or those who have discontinued use, but the directions of causality between drug use, other criminal behavior, and other age-related processes are still ambiguous.

There are also important questions about the way in which the criminal justice system affects the course of criminal careers. During the twentieth century, the principal strategies to prevent crime have been rehabilitation, deterrence, and incapacitation. Studies of prison-based rehabilitation have largely reported failure—whether because there are barriers to the use of theoretically sound techniques or because there are flaws in the techniques themselves is not always clear. And while deterrence is effective, research on the structure and operations of the criminal justice system shows that it is very difficult to increase deterrence much beyond present levels. Simply stiffening the penalties on the books does not translate into a tougher, more deterrent crime system at the operating level. More policing, including recruitment of citizen-watch groups, and additions to jail capacity, is helpful, but limited in extent.

Incapacitation is at present attracting substantial scholarly interest, and considerable research has tried to measure quantitatively the cost-effectiveness of imprisonment. While a jailed burglar cannot break into homes, jail is an expensive proposition. If a 22-year-old burglar whose criminal career is likely to terminate anyway in 5 years is imprisoned for 10 years, the last 5 years of jail occupancy may be "wasted." Considerations other than cost-effectiveness clearly matter a great deal in setting correctional policies, but in taking these kinds of calculations into account, it is very important to have precise data on criminal activity. Knowledge of the "normal" course of criminal careers may provide a useful tool for measuring with greater accuracy the social benefit of incapacitation.

Since criminal careers occur over time, longitudinal panel techniques are now likely to yield more powerful results than additional cross-sectional studies. Some of the main understanding of criminal career patterns came about from early panel studies that began in the 1930s. Additional studies are now needed to fill out in much greater detail the main points of the criminal life cycle: early childhood development in the family or institutional context; the onset of offending, which may be as early as middle childhood; the age of peak participation or prevalence in the mid to late teens; and the settling into and dropping out of criminal careers at later points in the life cycle.

Ideally, such studies might follow a single cohort panel from birth to age 25 or older. The problem with this approach, however, is that it takes decades to complete to get the full payoff in knowledge. An alternative approach is to select several cohort panels of different ages, following each of them for a period of several years to reach overlapping ages, and to ascertain the relative effects of criminal justice interventions. The complex design of such studies, based on interviews, observations, and records, now is being undertaken. Design includes the development of specialized institutional arrangements to collect, maintain, and govern access to the data and, when possible, to link such longitudinal designs to randomized field experiments involving shifts in policing or judicial operations, family-based or school-based interventions, or other changes that may affect criminal behavior (see "Experimental Design," Chapter 5). Additionally, comparative studies of crime and criminal careers in other countries and during other periods of history can provide a richer context and help overcome some interpretive limitations on cohort studies, enabling researchers to pinpoint which factors are specific to this particular period and which are more general to understanding the origins of criminal behavior.

Antisocial and Prosocial Dispositions

An area that needs to be examined much more systematically concerns the effects or criminal activity of such key life events as marriage, getting a job, joining the military, moving to a new community, and the like. A negative association has been shown between having a legitimate job and committing violent crimes (for adults), and there is evidence on the relationship between certain kinds of family background and a career in crime. The influence of family structure and dynamics during childhood on criminal behavior seems very important. In correlational studies based on regional statistics, communities with high divorce rates also have high rates of criminal activity. But these data cannot reveal whether it is actually the children of divorced parents in those communities who are responsible for the crimes. Children with only one parent because the other has died are less likely to be involved in offenses than children whose parents are divorced, which suggests that the existence of marital conflict surrounding divorce, rather than the fact of losing a parent, may matter most in this connection. Children of criminals are more likely to commit offenses than children of noncriminals. In recent years, there has been substantial research interest in the problem of family violence, but there is as yet insufficient knowledge of the way in which a history of child abuse or other forms of family violence may be connected to later criminal activity. Researchers now want to use more sophisticated techniques to identify the exact relationship between family conflict and later criminal activities, as well as the effect of strategies of intervention.

One likely factor in the development of antisocial behavior, by at least some adolescents, is how their parents typically try to manage their children's behavior. In one study of adolescent males, measures of parental behavior were categorized as monitoring the young man's whereabouts, disciplining him for infractions of various sorts, teaching him problem-solving skills, and reinforcing desired behavior. The adolescent's behavior was measured in the areas of delinquency (as indicated by police and self reports), academic performance, and social relations with peers. The study found that higher rates of the (intercorrelated) parental behaviors of monitoring and disciplining were associated with decreased delinquency; academic performance and social relations were strongly associated with parental teaching and reinforcement; and the other pairings showed little or no association. That these are probably causal connections was established by training the parents of chronic delinquents to do what other parents typically do about monitoring and discipline; the result was a significant long-term reduction in the delinquent behavior.

Another line of relevant research concerns the linkages between empathy, prosocial motivation, and aggressive behavior. Empathy is a class of emotional responses to someone else's situation rather than to one's own. Research has demonstrated that most people respond with empathic distress to someone in danger, discomfort, or pain. There is also evidence indicating that empathic distress is largely involuntary: it is difficult to avoid empathizing with someone in pain or distress without actively engaging in certain evasive perceptual or cognitive strategies, such as looking away from the victim or trying hard to think about other things.

It has been clearly demonstrated that empathic distress contributes to prosocial, helping behavior in most people. Such helping behavior is usually motivated internally rather than by the desire for reward or approval (although these motives may also be present); moreover, the occurrence and rapidity of helping behavior are related positively to the intensity of empathic distress. When people give help the intensity of their empathic affect is diminished. An as-yet small body of research suggests that empathy may also reduce unprovoked aggression in children and that empathy training can reduce the aggressive behavior of adolescent boys with delinquent tendencies.

The effects of empathy are, of course, limited. For example, the relationship between empathy and helping behavior is often mediated by whether the person in need is viewed as similar to the observer (empathic bias). But there is evidence that empathic bias may be reduced by educating people about what they have in common with members of other cultures or groups, as well as with other individuals in their own group.

The potential fields of application of empathy research include childrearing practices, the school setting in and out of the classroom, interactions with peers, and the content of television and other media. The challenge is to learn enough about how empathy develops and functions to permit advice to parents, media, and training programs on how to increase the empathic potential of delinquent or criminally prone youth (and others), whose sensitivity to other people may have been blunted by previous life experiences. Apart from this value, a better grasp on the conditioning of a human response tendency like empathy may illuminate the major cohesive, as well as divisive, forces in society.

ATTRIBUTIONS AND EXPECTANCIES IN SOCIAL INTERACTION

The concepts of self-esteem, social networks, and gender-based differences have been vital to research advances on emotional maturation, health risk and protection, and criminal behavior, discussed above. They have also been used to guide research on topics such as educational attainment, language development, and mental frameworks, a range of topics taken up in Chapter 1, as well as research on bargaining, employment relations, and fertility behavior, discussed in Chapters 3 and 4.

These theoretical concepts about social-psychological mechanisms have a familiar ring to them. The terms themselves seem part of today's universal vocabulary, and it is easy to assume that they are commonsense ideas that have been handed down from generations. But the terms, and the nuanced ideas they represent, were actually invented by behavioral and social scientists not long ago, tested and polished in classic theoretical and empirical work, and then became diffused (as are other scientific ideas) through the broader culture through education and the mass media. The concepts have since been adopted so fully, applied so widely, and their very novelty forgotten so rapidly as to seem truly everyday knowledge.

In the past, experimental research on interpersonal relations tended to focus on isolated components of what is an interactive process. In the interest of sufficient experimental control, participants were often strangers who were prevented even from directly seeing each other. More recently, however, the study of social interaction has shifted from the identification of particular factors implicated in certain products of social interaction—a happy or unhappy marriage, a fast or slow work group, a successful or unsuccessful negotiation—to an understanding of the process. The emerging paradigm is the simultaneous study of many elements, freely varying in naturalistic settings, assessed over long durations, involving participants who have both a joint history and an anticipated future together, and using multiple methods such as self-reports, observer ratings, behavioral codings, and psychophysiological measurements.

This change in the study of social interaction has been possible because of developments that include technological advances, such as high-fidelity videotape, psychophysiological recording, and powerful microcomputers; advances in statistics, including new techniques for time-series analysis, model estimation, and refinements in multiple regression and analysis of variance; and advances in theory formulation, including computer simulation as a medium for writing more formal, quantitative theories. As a result, researchers are now beginning to make substantial new inroads on such questions as when erroneous expectancies (for example, stereotypes about race or gender) are most likely to be perpetuated, which interactive features improve the efficiency of task groups, or what principles govern the growth of initial encounters to close relationships.

Expectancies, Self-Concepts, and Motives

In an important experiment in the 1960s, some teachers were led to expect that certain of their 1st- to 6th-grade students would "bloom" academically.

Although the designated students had been randomly chosen, subsequent testing following a period of instruction showed they in fact did better than students who had not been designated in that way. Although this study aroused considerable controversy and was criticized on a variety of methodological grounds, subsequent research has generally supported its conclusions. The findings were important not merely in showing that teachers were able to influence students, but also because the teachers, operating with an incorrect hypothesis, behaved in such a way as to bring about the confirmation of the hypothesis. And the teachers were not just misled by the false expectancy to misjudge the performance of the designated students; they created a situation in which the students indeed did perform better by independent, objective measures, and the teachers were unaware that they had done anything special to create this state of affairs. In its general form this concept is known as a "self-fulfilling prophecy."

Research on self-fulfilling prophecies shows the manner in which features of personality, at least in terms of how a person appears to outside observers, develop and change over time through social interaction. Rather than being preprogrammed and static, like a phonograph record, personality is, to a considerable degree, constructed and reconstructed through relations with others. This phenomenon has been clearly shown in studies of aggressive boys. This work identifies a process in which these boys expect aggression from their peers and accordingly interpret ambiguous provocations as reflecting definitely hostile intentions rather than seeing them as neutral probes or as possibly accidental. This leads the boys to retaliate aggressively, a reaction that confirms their reputations for aggressiveness and elicits peer counterreactions that further strengthen the boys' beliefs about the hostility of those around them. Unfortunately, however, this work has not yet discovered how such cycles begin. An important advance along these lines would be longitudinal research that permits the separation of intrapersonal and interpersonal factors.

The subject of motives and how they are internalized is also an active area of current research. It has been shown that providing auxiliary rewards for performance of an initially attractive activity seems to rob the activity of some of its intrinsic interest. In a classic controlled experiment, for example, children showed a decline in their interest in playing with "colored markers" in a free-play period that took place some time after they had been rewarded for playing with them. Subsequent experiments have shown that this decline in intrinsic motivation does *not* occur if the reward is given as a sign of competence or excellent performance; in fact, intrinsic motivation can be enhanced under these circumstances. The effects of reward structure on intrinsic motivation are also mediated by differences among individuals in their motivations to be competent.

A particularly important feature of human social interaction is that people intuitively understand many of the principles governing social processes and, as a consequence, are capable of manipulating them for their personal purposes. Research of the last two decades has examined the skills that people use in presenting their personalities to other people. For example, for some people whose self-image is competent, if that self-image is threatened, then they will try to protect it by handicapping themselves in ways made available by the environment, so that they have an excuse for poor performance if it occurs and gain credit from good performance if that occurs. In the case of strategic self-presentation, in which the personality presented to others is not in line with the person's self-concept, the attempt at social deception can be unmasked by subtle features of facial expression. A major line of research in social interaction focuses on the microprocesses through which information is communicated nonverbally, and sometimes unconsciously, about personality.

Development of Close Relationships

After years of studies restricted to individuals in isolation or to relationships between (usually) two people who had never seen each other before and probably never will again, research on relationships is moving in new directions. Changes in the internal and interpersonal processes that accompany the emergence of an enduring relationship are now being seen in laboratory studies of interaction between people who expect that their initial encounters will be followed by more lengthy ones. The earlier research had shown that, in situations where experimental interactions between subjects were entirely casual and without any long-term consequences, people tend to rely on stereotypes and first impressions. More recent research has shown that even the possibility of additional future interaction has dramatic effects on individual behavior, both before the first contact takes place and during its initial stages.

The anticipation of future interactions appears not only to heighten attention to the other person, but it also to contribute to the development of a more individuated impression of that person. For example, one recent study shows that a person anticipating future interaction with someone else pays more attention to information inconsistent with his or her preconceptions about the other person and tends to process that information carefully for its implications about the other's traits and attitudes. In contrast to tendencies to discount such information, as in the intrapersonal processes supporting self-fulfilling prophecies, the person who anticipates a future relationship frequently relies on unexpected information to revise his or her initial impressions.

In laboratory studies, the possibility of a future with another person changes the format of interaction regarded as most appropriate from one of "social exchange," in which a benefit is given in response to receipt of a benefit, to one of "social responsiveness," in which the benefits given take into account the perceived needs of the other person. The details of interactional differences under these two circumstances are being explored; people sharply distinguish the two kinds of relationships and have characteristic ways of expressing their different levels of responsiveness. The shifts in motivation as relationships develop are captured in recent theories that analyze systematically the initial patterns of incentives and the formal transformations that may be performed on them.

Some measures of the changing nature and quality of social interactions over extended periods of time are provided by research on shifts in levels of satisfaction over the course of an enduring marriage or career. For example, marital satisfaction is typically high early in marriage, decreases with the birth of the first child, reaches its nadir as the children enter adolescence, and increases as children leave the household. Work satisfaction increases until approximately age 40, levels off through the mid-50s, and rises again thereafter. There are a number of competing explanations for these trends. The investigation of spousal, parent-child, coworker, and worker-supervisor relationships at different phases in the life cycle is the next step in understanding social interaction processes as they evolve over long periods of time.

Small Groups and Behavior

Frequently, large organizations such as governments, corporations, or universities create small, temporary groups to carry out specific tasks. The capacity to form, activate, and dissolve such ad hoc committees or working parties is a major adaptive resource of organizations and societies and has become a frontier area for research on decision making and other fundamental group processes.

Some of the research conducted on ad hoc groups is designed to examine the scope and variation of basic propositions concerning group influences on behavior. One important instance is the following effect: when an individual is asked in experiments to perform simple but physically demanding tasks, he or she typically exerts substantially less effort when led to think that a group or people is doing the tasks at the same time than when led to think he or she is performing them alone. While the total work done increases steadily as the actual number of people in the group increases, the average effort exerted by each individual decreases steadily with group size. This type of research provides a baseline for studying the conditions under which various kinds of incentives induce greater or lesser individual productivity: How much does the effect obtain when the labor involved is mental rather than physical? When the group is comprised of friends rather than strangers? When each individual's task is unique? And what is the trade-off between greater individual effort and greater need for coordinating differentiated tasks? Research on these questions is obviously important for understanding productive efficiency in many settings.

In a different vein, a series of experiments brought together strangers who

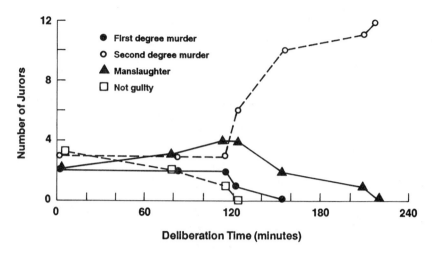

Deliberation Time (minutes)

JURY DECISION MAKING   How do trial juries reach decisions? What is the effect of different agendas for making a decision, of changing jury sizes, of varying decision rules—for example, unanimity versus majority?

Among the new experimental methods for studying collective decision making is the simulated jury trial, in which a group of people are selected from actual jury panels to watch a videotaped trial proceeding and then, in an actual jury room, deliberate and reach a verdict while their discussions are recorded. This figure illustrates the waxing and waning of groups of jury members speaking or voting for particular verdicts across the four-hour deliberation of one such mock jury deciding a murder trial.

The pattern displayed in this 12-member, unanimity-rule jury deliberation is associated with the most common of several observed decision agendas: "presumed innocent." The jury first addressed basic evidentiary issues: whether the defendant was guilty of any of the charges. During this period, the members were evenly split—in a typical pattern—among five possibilities— not guilty, manslaughter, second-degree murder, first-degree murder, and undecided. During this "evidence-driven" phase, few shifts of opinion occur concerning the proper verdict. In a series of votes after about two hours, the jury decided that "not guilty" was untenable, and the debate shifted directly to the verdict. Faction sizes then began to shift dramatically. The final verdict, second-degree murder, was the modal one for all mock juries viewing this trial. It was also the one favored by most judges and attorneys, and it was the verdict of the original jury at trial.

The pattern shown in this figure, and the probability of reaching a particular verdict, would differ if the size of the jury were smaller, the decision required a certain majority but not unanimity, or the jury adopted a different discussion agenda, such as "murder (first or second degree) versus non-murder (manslaughter or not guilty)."

differed noticeably with respect to race, age, sex, social class, and language use—attributes differentially valued in the society. The people were presented a series of perceptual problems to be solved (for example, does a rectangle enclosing smaller black and white rectangles contain more black area or white area?) for which their status characteristics were objectively irrelevant. In the absence of prior information about one another's task-relevant skills and knowledge, persons with high-status characteristics were often attributed more of the required skills than were low-status persons. Consequently, they were allowed to initiate more interactions, and their solutions were more readily accepted by the group. In short, the status ordering produced within the group reproduced the external social order rather than responding to the problem at hand.

An important and revealing group process occurs in trial juries, which vary from one jurisdiction to another in size and in the proportion of assent required to assert a verdict. How do differences in size and rules—say, a unanimity rule in a jury of 12 compared with a 6-person majority in a jury of 8—affect the decision process and outcome? One can, of course, speculate, and jurists have, but there has been little empirical knowledge to support these speculations. What goes on in a jury room has not been subject to close examination except by later reconstruction of the proceedings from the memories of jurors, which are known to be fallible at the level of detail needed. However, carefully developed studies of experimental juries have now yielded powerful new knowledge. In an example of such experiments, several panels of experimental jurors, selected from actual jury pools, watched a videotaped enactment of a real trial, and then met in regular jury rooms to decide on a verdict. Among other things, the group size and decision rule employed were varied from panel to panel. The proceedings were recorded on videotape (with the jurors' knowledge) and then subjected to intensive computer-aided analysis. One major finding concerns different decision rules. Researchers found that when unanimity is required, the deliberations are far more thorough than when it is not required. A large fraction of the additional discussion, uncovering and eliminating serious errors of fact and law, occurs after a decision would have already been reached under a plurality rule. Moreover, nonunanimous juries more often reach extreme verdicts (for example, first-degree murder rather than second-degree murder) than unanimous ones do. With a nonunanimous rule, members of small factions also contribute less to discussion, and larger factions attract new members more quickly. This finding is particularly significant because in other contexts it has been shown that the outcome of a group discussion is determined by those members who frequently shift their opinions, rather than by those with more extreme initial positions.

These findings focus attention on the dynamic interactions among group members rather than on dependencies between pairs of variables measured at only two points in time. They have led to the development of more complete theoretical accounts of the dynamics of interaction in groups. They have also been cited by legal policy makers concerned with constitutional guarantees of due process and the selection of representative juries. Such research on group decision making exemplifies the theoretical and practical contributions from research on social interaction.

The Social Construction of Gender

It is becoming increasingly clear that human sexual differences are actually composites of variable elements. There are chromosomal differences (for female and for male mammals), anatomical differences, physiological differences, psychological differences, behavioral differences, and sociocultural differences. The latter three categories are generally studied under the rubric of gender research, with several components:

- primary gender identity (male or female),

- partner choices (heterosexual, homosexual),

- gender-relevant behavior styles (female "tomboyishness," male "effeminacy"), and

- sexually dimorphic nonsexual capacities (for example, geometric ability).

Recognition that gender can be treated as a behavioral and social construct quite distinct (though not necessarily divorced) from biological differences has led to important advances in a number of fields.

New Guinea has proven to be something of an anthropological laboratory on gender differences. For decades, this island was known to house cultures supplying some of the world's most exotic gender beliefs and practices: ritualized homosexuality, elaborate notions of menstrual pollution, ceremonies of sex-role reversal, and the most extreme known doctrines of male supremacy. Analysts relying on earlier theories were unable to explain or interpret these unusual beliefs and practices.

More recently, however, researchers working with social-construction theories have begun to unravel a consistent native logic underlying these exotic data, detailing the way in which New Guinea peoples use food, sexual activities, and ritually concocted substances to think about and manipulate kinship and gender identities, health and disease, life and death. In some areas of New Guinea, for example, boys are ritually "grown" into men by feeding them male-grown or male-hunted foods; in other areas, by putting semen on or in their bodies; and in yet other areas, by ritually bleeding them to rid them of "female blood." In every case, a logic comes to the surface concerning social identities that can be manipulated because they are bound up with substances that can be manipulated. Witchcraft beliefs and notions of kinship have also been shown to link consistently with this logic. Furthermore, and more surprisingly, regional variations—for example, between an emphasis on adding male substances and an emphasis on deleting female ones—have been shown to covary systematically with variations in political organization between different tribal groups.

Researchers trying to understand social change in Europe have also begun to adopt a social-constructionist approach to gender systems. Western history is not without sexual exoticisms—the nineteenth century has attained a status among social historians similar to New Guinea in anthropology. The Victorian evidence shows the extent to which notions of gender may be reconstructed in response both to received ideas and to emerging social, economic, and political patterns. For example, outbursts of peculiar sexual theorizing (say, that masturbation causes insanity or that sexual restraint is analogous to capital accumulation) have been shown to be related to stressful changes in family structure and class relations.

The interpretive methods leading to these breakthroughs in understanding gender systems in geographically and historically diverse locales are just beginning to be turned toward understanding gender systems closer to home. A growing focus is the complex interrelationship between what goes on in the public place of work and the private place of the family. One of the most challenging questions is to explain the motivational force and historical resilience of gender beliefs: for example, that "women's place" is in the home or working as a nurse or elementary teacher, which draws upon their "natural" capabilities, while men are "better suited" for work that requires highly rational thought or the exercise of authority. To the degree that earlier research on gender behavior was heavily influenced by such traditional models of male and female behavior, it tended to leave unstudied a large range of behavior that did not fit the stereotypes, such as the diverse activities of women in the public domain and that of men in the domestic or private domain. Over the last decade, under the impetus of new theory and its implications, researchers have compiled a much improved record of the public and political activities of women: biographies of hitherto little-known activists; data on women's labor organizations and political action in the workplace; data on women's rights movements at different points in American history; and descriptions of reform movements directed at broad social issues—temperance, slavery, and peace. Research on men's domestic activities lags substantially behind this work, but when more data are uncovered, one can anticipate a thorough sorting out of competing theories and stereotypes in light of a more complete historical record.

OPPORTUNITIES AND NEEDS

The anatomy and control of motivational processes; the codification of emotional expression; and the understanding of eating, sexual, and other behavior in animals and humans have all been illuminated in recent years by continuously improved technology, better controlled research designs, and more sophisticated theoretical ideas. Research has given insights on the role of peer pressure, mass media, and market forces in changing societal levels of alcohol, tobacco, and drug abuse. Controlled policy experiments have revealed effective strategies for designing medical cost-control measures without detrimental effects on health. The career dimensions of criminal activity have been opened to study, and important lessons have been developed for crime prevention and control. The study of small group processes has been enriched by innovative experimental approaches such as simulated jury trials facilitated by video recording equipment. The subtlety and deep-seated effects of gender differences and divisions have been clarified by sustained cross-cultural and historical investigation. Research in these areas often involves complex interactions with large, well-established professional groups and organizations: for studies of biobehavioral aspects of health, health care providers, physicians, and hospitals; for studies of criminals, the police, the judiciary, and the penal system; for realistic studies of juries and the courts. In each case, complex questions of access, confidentiality, legality, and skepticism must be answered before any study can even begin. In the case of longitudinal studies, these issues must be repeatedly confronted, often in the context of changing institutions and technologies. We therefore propose new expenditures of approximately $56 million annually to make possible further developments in this research.

One of the outstanding needs is to renew the technological bases for laboratory and field research, for which we estimate about $10 million per year is needed. The tools of neuroscience, especially in the realm of surgical and physiological equipment and neuroimaging devices, have undergone revolutionary changes in precision, skill requirements, and expense. New generations of audio and video recording and synthesizing equipment permit more elegant research designs and aural and visual displays than were possible when experimental stimuli and theoretical models had to be constructed, presented, and controlled by hand or with mechanical or photographic media. Recent improvements in physiological measurement technology using biochemical and endocrinological assay techniques now make it possible to simultaneously examine behavioral and biological states and responses and to do so in social settings such as homes or workplace. Recent statistical innovations permit investigators of interaction and communication to examine reciprocal causation over time more readily and precisely than in the past. At the same time, advances in video technology permit more naturalistic social interactions to be recorded cheaply and preserved indefinitely so that the qualities of interpersonal relationships can be sampled over long periods, which is increasingly important as more investigators move from the study of first impressions to the study of longer-term close relationships. Finally, the revolution in microelectronics has made mainframe computing capacity available in desktop or mini machines, affecting practically every kind of research.

While new equipment for behavioral and physiological measurement and analysis is much more powerful and flexible than that available only a few years ago, budget limitations on equipment acquisition and methodological training have left many laboratories and field stations in a state of actual or potential obsolescence. The laboratory facilities in most university departments were developed and outfitted before the advent of modern video, computer, and other technologies. The present stock and potential demand for modernization of laboratory equipment needs to be systematically canvassed, especially in multiuser laboratories, to determine an appropriate set of goals and schedules for upgrading. Provision for these facilities should also include technical support, training in the methods and analytical techniques appropriate to new technologies, and graduate and postdoctoral research opportunities, discussed below. Of the $10 million more annually that we estimate is needed for technological upgrading, we estimate that $5 million should be added to laboratory equipment expenditures, $2 million more specifically for neuroimaging technology acquisition and access, $2 million for computer hardware and for software development, and $1 million for upgrading of research animal care.

The need for intensive technical training in the operating skills, underlying principles, and new theoretical possibilities linked to the technological advances requires an infusion of support for research fellowships, traineeships, and institutes. Overall, we believe that $8 million annually should be added for this training. At the graduate level, the need is not to increase the overall number of graduate students, but to reduce their heavy reliance on teaching and outside income during these years, so that as new doctoral scientists they will have research experience well beyond the boundaries of their own dissertation topics. At the postdoctoral level, the need for additional training opportunities should be more closely integrated with the importance of building and sustaining collaborative affiliations and countering the fragmenting effects on science of rapid growth in knowledge and shifts in technique. A sharp increase in fellowship support is needed to achieve the kinds of technical skills required in some of these areas, and leads to a recommendation of an increment of $5 million, with the majority ($3 million) allocated to postdoctoral support and the minority ($2 million) to predoctoral support. We also recommend that $3 million be committed to advanced training institutes.

The value of longitudinal studies—including those with experimental components—has been thoroughly proven by outstanding work in each of the research areas discussed in this chapter. Prospective longitudinal studies are usually the most effective way to uncover and generate certainty about causal relationship with extended temporal or developmental structures. But the time between their design and the reaping of their results can be as much as from 20 to 30 years, during which time there are inevitable improvements in technique and changes in the scientifically and practically significant questions. To some degree an ongoing study can accommodate changes, but some shifts in knowledge and interest will be of sufficient magnitude to warrant beginning new studies on a topic before older ones are completed.

The past decade has seen a decline in the rate of new longitudinal studies, cutbacks in the frequency and scope of data collection in several mature studies, and high rates of inactivation in both new and mature projects. It is time to revitalize this critical aspect of the research system. Commitments are needed for longitudinal studies of normal affective and motivational patterns, emotional maturation, and the incidence and effectiveness of treatment across the lifespan for affective illness; risk factors and preventive interventions (including marketplace variables) with respect to alcohol, tobacco, and drug use and sexual and dietary patterns; the role of social networks in affecting health, particularly later in life; the development of criminal behavior in the critical period from middle childhood to young adulthood, especially in the context of family behavior and criminal justice policies; the development of close personal relationships; and the development of changes in gender roles and relationships. For this substantial agenda we recommend a new commitment of $15 million annually, which will permit approximately 10 large-scale longitudinal studies to be carried out.

It is essential that the multidisciplinary character of research on motivation, behavior, and social contexts be recognized and appropriately supported by research agencies. To do so will require some reconsideration of relevant review processes, broadening of administrative protocols to encourage collaborative research efforts, and review of staffing for extramural research. In particular, one of the major routes for advancing collaborative research is the creation of various types of research workshops. They can include brief, recurrent meetings of a group of active collaborators and extended workshops (4–6 weeks) at which research is reviewed, planned, and carried out. For these purposes, we recommend an annual increment of $1 million. The staffing requirements of support programs that try to encourage, properly review, and adequately monitor collaborative and multidisciplinary work are more demanding than single-investigator review and funding operations. We strongly encourage modest increases in the numbers of program staff in funding agencies so as to permit greater attention to solving the problems of integrating research portfolios, encouraging grantee interchange and collaboration where appropriate, and cultivating a broader knowledge base about the scientific and administrative opportunities, benefits, as well as pitfalls of collaborative or at least more reciprocally informed and mutually responsive research.

Of particular concern is the segmentation of research on health matters according to disease-specific missions. This segmentation tends to inhibit the basic study of common biobehavioral and sociobehavioral processes and, indeed, of health-related behavioral and social processes on the whole. There is a need across the health research institutes in the National Institutes of Health (NIH) and the Alcohol, Drug Abuse, and Mental Health Administration

(ADAMHA) for increased coordinated research on common biobehavioral processes relevant to health. Such projects need to be reviewed by interdisciplinary research specialists; the skills and perspective of such specialists cannot be simulated simply by organizing review panels with researchers from different disciplines. In addition, at the National Science Foundation (NSF), review practices discouraging the support of studies with clinical samples should be reconsidered. While NIH devotes substantial efforts to support of fundamental biological research that involves clinical subjects, this is not true of behavioral and social sciences research support from NIH. It would not be duplicative for NSF to increase its investment in such research.

There is a growing sense among researchers on affect and motivation that the establishment of national multidisciplinary research centers, in which teams of investigators can combine diverse approaches and methods, is one of the best means of advancing research on these processes. Such centers could bring together researchers who do not now communicate much, particularly those concerned with normal affective and motivational processes (principally psychologists) and those concerned with the causes and treatment of affective and motivational disorders (principally psychiatrists and neurologists). An annual expenditure rate for such centers of as much as $9 million is recommended.

Finally—and in spite of the emphasis given above to collaborative research and centers—we recognize that much of the work in all of these areas is advanced through individual investigator grants. Even with refurbished laboratories, better trained researchers, and high-quality data bases, there is still a need for grant support that is tailored to the specific work and interests of qualified researchers, and such support must be adjusted to take account of the more complex requirements and longer time horizons of present research frontiers. We recommend a steady increase for individual grants over the next several years to appropriately $13 million more than present funding levels. Most of the increase should go to increasing the size of individual grants to more realistic levels for the work; the balance should be used to fund additional investigators, especially young ones who are currently being denied funding even though their proposals receive high ratings in the review process.

3

Choice and Allocation

3

Choice and Allocation

This chapter is mainly about situations in which people make choices and goods and services are distributed. The most familiar such situation is the market, but choices and allocation play essential roles in all organizational and political contexts. Among the questions that have come to dominate recent research in this area are the following: What are the distinctive features of *collective*—in contrast with individual—choice and decision making? Is it true that "who controls the agenda controls the decision," and if so, in what sense? What do electorates really choose, and how does the choice situation presented to them affect the outcome? What are the forms and consequences of internal political struggles within organizations? How do external, institutional constraints (such as constituencies interested in the fate of organizations) affect these processes?

In looking more narrowly at markets and related economic activities, current research considers such questions as: How are choices in markets affected by the availability of information and the structure of incentives? When and how do efforts to influence incentives through regulation contribute to more or less efficient market systems? To what extent are the expectations of economic agents based on "rational" factors and to what extent are they influenced by other factors?

What considerations do economic agents take into account when they negotiate and strike bargains with one another? How exactly do repeated interactions and stability of a relationship among agents change these processes?

How do complex patterns of information, incentives, constraints, and discrimination affect wage rates and other outcomes in the labor market?

The issues of social groups selecting among alternative possibilities and deciding on allocations of scarce goods and resources, coupled with the use of power to enforce certain decisions, are central to two disciplines, political science and economics. But other kinds of knowledge are also involved. Since organizations are often the key to carrying out those choices and allocations, insights from social psychology and sociology are needed, and recent work has incorporated laboratory experimental methods evolved in behavioral psychology. Many issues are addressed most effectively by combining all of these perspectives.

COLLECTIVE CHOICE AND ORGANIZATIONAL BEHAVIOR

Decisions are inevitably affected by faulty memory, limited capacity to process information, and uncertainty about various factors that affect the outcome. Within these inescapable constraints, people who must make decisions should take into account all of the possible consequences of choice, good and bad, and their probabilities of occurring, not overlooking events whose probabilities are extremely small. The growth and understanding of individual decision making was discussed in Chapter 1. But important decisions are often assigned to groups—boards of directors, committees, legislatures, and the like. Such decisions, or collective choices, involve a social process that introduces a range of considerations quite different from those involved in individual decision making.

Research on collective choice has focused mostly on formal decision mechanisms of voting and on resource allocation, especially of public goods. Of particular interest is a mathematical approach that illuminates and yields strong predictive power in analyzing legislative agenda formation and electoral procedures. A major challenge at present is to understand the group processes that underlie many private decisions in the business world as well as issues that arise in public voting. This work on group process, carried out mainly by investigators in social psychology and organizational behavior, has amassed a great deal of data and has led to a considerable body of informal theory, but so far detailed formal or computer-simulated models have been rare. Work in this area is expected to grow, in part because of the enormous economic and social importance of improved decision making to both government and business.

Setting Agendas

Many organizations have the capacity to make certain choices that are likely to be disadvantageous or even oppressive to some members, and even in the face of dissent or dissatisfaction they can, to a degree, enforce these choices. At the highest level of social organization, taxation without reciprocal services, conscription to fight in a war that some individuals, alone or as part of an organized minority, regard as illegitimate, and imprisonment are extreme examples. Most organizations, even fundamentally political ones, are formally voluntary—citizens can, in principle, leave the city, state, or nation; stockholders can sell their stock; workers can quit their unions—but these "exit" options may be so costly, unattractive, extreme, or ineffective that they are hardly real options. More serious are the options of vocal contention, disruption, or withdrawal of support, active participation, enthusiasm, and diligence. The need to maintain such support in spite of disagreements and conflicts with members leads to the establishment of complex and sophisticated procedures for legitimating decisions, generating loyalty, and resolving intraorganizational disputes.

Because majority-rule voting systems are explicit, formal, and very common, they are the best understood of such procedures. A centrally important, undesirable feature of such voting systems is that they typically exhibit a particular kind of fundamental indeterminacy. For example, suppose that a group of three people (or three voting factions of equal size) uses a majority-rule voting procedure to select one of three alternatives— a, b, or c. Suppose further that the first voter or group prefers a to b and b to c, and so a to c; the second voter, b to c and c to a, and so b to a; and the third voter prefers c to a and a to b, and so c to b. Thus, one simple majority (voters 1 and 3) prefers a to b, another majority (voters 1 and 2) prefers b to c, and a third majority (voters 2 and 3) prefers c to a. No alternative is preferred to both of the other two. So the choice of any one of the three alternatives appears to be wholly arbitrary and will be determined by the order in which the alternatives are considered.

A major theoretical finding of the 1950s, whose significance is only gradually being recognized in practice, was a general proof that whenever a system of majority rule is used to choose among more than two alternatives, the outcome will in general depend on the order in which pairs of options are considered, the so-called voting agenda. This general mathematical property of majority rule under these conditions makes it highly susceptible to manipulation. It was shown, under plausible assumptions about the distribution of individual preferences, that if the agenda for voting can have any effect at all, then it completely dominates all other effects. The voting order can be selected so as to lead a group to choose nearly any option on the table, including those options that virtually everyone would initially consider to be undesirable. The logical structure of majority rule thus entails the possibility that anything can happen, depending on the agenda sequence. This means that control over the agenda— over the order of voting or other procedures—is extraordinarily important, as is obvious in the simple example above.

These theoretical observations have been shown to apply empirically to many complex and realistic situations. Whenever the number of alternatives is sizable, the probability that there is a single, universal majority winner over all other options is very small, and the outcome depends largely on the procedures for determining the order of consideration of the alternatives—the agenda. Recent experimental work clearly demonstrates that an agenda of a particular form, implemented and rigidly followed in situations in which individuals are unaware of other's preferences (as in secret-ballot elections), can succeed in controlling a group decision. Furthermore, this phenomenon has been shown to hold not only for majority rule, but for most of the voting rules in common use that entail subdividing the set of alternatives into a series of votes.

Two avenues of research in this area are now under intensive study: explicit treatments of improved, fairer agenda setting for those social decisions that involve sequential voting among alternatives, and the development of amalgamated-preference decision schemes that require only a single, simultaneous vote and are not subject to serious distortions due to strategic voting. One important approach in this area, approval voting, is discussed below.

Sequential and Simultaneous Votes

Since the source of power in organizations rests heavily on the ability to affect agendas, both their constitutions (or other fundamental contracts) and the strategic behavior of conflicting parties (for example, their rhetorical presentation of issues) must be analyzed in terms of agenda processes. Basic theoretical research on agendas begins with the much simplified case in which agenda items arise in a known order and the preferences of all participants are fully known to everyone. As research knowledge about such extreme cases has increased, it has become possible to study rigorously cases much closer to the real world, in which knowledge of the preferences of others is imperfect. Ultimately the test of any attempt to design a better agenda-setting procedure is that it should both be easy to implement and should reduce the probability of disastrous outcomes, such as electing candidates who are low on everyone's ranking or those who are backed by an intense, organized minority even though these candidates are disapproved by the majority.

The natural alternative to sequential voting procedures are ones in which each voter must report, at one time, something about his or her preference ordering of all of the alternatives, and this information is amalgamated by some specific rule to generate the social choice. A famous result, much studied and elaborated, shows that if all voters provide a detailed ranking of their alternatives, it is mathematically impossible to devise a fully satisfactory procedure, one that meets all of the usual requirements of fairness, to yield an amalgamated group ranking. The most pernicious feature of most procedures for amalgamating preference orderings is that they invite strategic voting, in which a voter reports deceptively about alternatives other than his or her most preferred one.

For example, in one widely used by small committees attempting to rank several alternatives, each voter ranks the alternatives from best to worst, and the group ranking is obtained by adding the numerical rankings assigned to each alternative. A common strategic move is for a voter to put his or her second (third, and so on) choices at the bottom of the list if those options appear to be the first choices of a substantial number of other voters. The effect of such a move is to increase the likelihood that the strategic voter's first choice will win out despite the fact that more people prefer one of the other choices.

Such strategic voting clearly creates a distortion in the social process, and realization of that fact has prompted a search for procedures that are far less susceptible to such manipulation, although they will necessarily have some other undesirable feature. One example, not yet fully understood, is approval voting, in which each voter partitions all of the candidates into only two sets, approved and not approved. The procedure seems to work well in practice if each voter splits the approvals and nonapprovals about equally, but additional theoretical, field, and experimental work is needed to understand more fully its properties—both limitations and virtues—when the approval/nonapproval split is not equal. To some degree, progress has been hampered by limited resources since the experimental and observational work involved must be quite extensive. But because voting is such a pervasive feature of modern societies, especially ours, it seems to be a wise investment to understand better how best to carry it out.

Electorates

One problem of institutional design that is important for contemporary society is to render electoral systems fair to individuals and to significant groups during periods of social change as well as during more stable times. In the United States, Great Britain, Australia, and Canada—unlike continental Europe and most other areas of the world—the most common election procedure is plurality: the candidate with the most votes wins. While single-member-district plurality elections predominate at the national and state level in the United States, multimember-district plurality elections are quite common at the municipal or county level. Plurality multimember districts have come under increasing attack in federal courts for diluting the voting strength of racial and linguistic groups, and testimony based on collective choice theory has played a critical role in these challenges to multimember districts. Certain provisions for majority runoff elections have also come under challenge as racially discriminatory in their effects. A body of research has been conducted during the past decade on the properties of runoff systems, and future work to fuse this analytic theory with other approaches may lead to definitive knowledge about these systems.

No country in the world makes greater use of balloting of one sort or another as a decision mechanism than does the United States, which has by far the highest ratio of elected officials per citizen. Only rarely is the choice between just two alternatives to be decided by majority. Not only are there often more than two choices, but much balloting rests on a complex federal system whose multitiered structure also generates layers in political parties. In addition, special majorities are required for certain kinds of actions; concurrence of more than one voting body is usually required for legislation (bicameralism); veto powers of various sorts govern the relationships between legislature and executive; one unit may have the power to propose and another the power to block; and so on.

Understanding the properties of such complex, layered organizational arrangements is no easy task, but much progress has been made in the past two decades. In particular, models based in part on game theory have permitted researchers to reexamine and develop new understanding about many complex institutional arrangements (for example, the veto powers of the "big five" in the United Nations Security Council, the character of the U.S. presidential electoral college, voting rules in the European Economic Community) in terms of a common framework. Such models have made it possible to discover when apparently minor changes in procedures may actually have major effects that otherwise would not be anticipated.

Research has also illuminated the link between types of electoral systems, distribution of partisan support (as measured in votes received by a party's candidates), and legislative seat shares. For example, an important hypothesis, the algebraic cube law of the relationship between the proportion of votes cast for a party across all districts and the proportion of seats it will win in the legislature, has been reformulated in a very general fashion, incorporating factors such as the average number of seats being contested per district and the effective number of political parties contesting them. This revised model permits far more accurate predictions than were previously available of the probable consequences of changes in election procedures, as in France's shifts between plurality and proportional representation. A number of key issues, such as changes in candidate and voter behavior in response to changes in election rules, await additional observations and theory to be incorporated into the model.

Going beyond matters of particular decision making to the overall theory of democracy, one central and long-standing problem concerns the relation between voting and the ideals of popular participation and control over government. In one popular vision of democracy, voting is expected to produce programmatically coherent results so that popular participation is sensible and effective. Yet analyses of multiple-option voting agendas reveals that it is quite possible for voting results to be persistently incoherent. This discovery challenges the town meeting view of the fundamental basis of democracy. In particular, the concept of direct voting as a means to enact directly a consistent popular will into law, which is rapidly becoming technically feasible because of electronic networks, simply may not be a coherent way to legislate. In contrast, the concept of voting as a means of changing officials and thus affecting the law at a greater remove—embodied in most U.S. legislative procedures—seems quite consistent with the discoveries of collective choice theory.

Founding Political Systems

A final major area of research in collective choice is the founding of political systems. Virtually all societies experience periods when the fundamental procedures of voting, agenda control, and dispute resolution become matters of conscious collective choice, perhaps exercised through an ad hoc representative body such as a constitutional assembly, a party convention, or a committee convened under military auspices and subject to plebiscitary ratification. The adoption of the Constitution is the most familiar example in the United States, but there are many others in recent history, including Japan, West Germany, and the numerous states that emerged from the territories of European colonies after World War II. At these moments in history, society (or at least parts of it) becomes the designer of its own organizational structures, and its choices determine in large part the future viability, effectiveness, and justice of those structures. Continued historical and comparative analyses of the creation of such political organizations—including studies of the alternatives then available—will lead to understanding the information, incentives, and conflicts that were pertinent to those who created them, as well as the effects of the procedures by which those organizations were created. It is also important to understand how later generations of leaders and citizens have read and interpreted those earlier historical moments and how they brought them to bear—successfully or not—on the economic, social, and political developments that could not have been envisioned by the organization's original designers.

Although overall research progress in the study of agendas and voting systems has been driven largely by theoretical work, a substantial commitment exists to empirical testing, observation, and application. The descriptive literature on political parties, interest groups, committees, and related organizational forms is rich, but needs further codification in terms of collective-choice models. Progress has been made in developing laboratory and systematic observational methods for studying collective choice processes, but this work is in a relatively early stage and will benefit from additional, focused research.

ORGANIZATIONAL DESIGN AND CHANGE

The past 20 years have seen considerable progress in research on the determinants of organizational structure. The first phase in this program of research developed what has come to be known as contingency theory. According to this perspective, optimal organizational design buffers the technological core, which is the material process of production, from external shocks. It does so by creating peripheral structures designed to deflect or absorb such environmental turbulence as market volatility, political change, major legal rulings, and the like. The optimal design depends on the detailed needs of the technical production system and the nature of the environmental variations and uncertainties.

In this model, if the technology is relatively stable and the environment varies along a limited spectrum of possibilities, the needed organizational structure is highly routinized and unchanging. Most organizational change is thus contingent on fairly revolutionary shifts in either the technological base or in the economic, political, or legal environment. More recent research has broadened the theoretically admissible sources of change—and sources of stability in the face of pressures for changes—by focusing on less formalized variables in the organization and its environment and studying closely the evolutionary movements that bring organizations more slowly but just as surely into new alignments with new capabilities.

Organizational Politics and Institutional Constraints

Research in the past decade has focused on two informal features of organization that have far-reaching implications: organizational politics and institutional constraints. Resource allocation within organizations is subject to intense political contest among agents within the organization, in business as in government. There are two main causes of such struggles. First, in large modern firms information and decision making is in fact decentralized, whatever the formal structure, due to the very limited capacities of decision makers, even aided by large computers, to observe, process, and communicate information efficiently. The largest private employer in the United States, General Motors, employs some 660,000 people, enough to fill every job in metropolitan San Diego or in the state of West Virginia. While most firms, even in the Fortune 500, are much smaller than General Motors, the median size of the these firms is still 13,000 employees, far too large an internal economy to be managed effectively without substantially decentralized information and decision making.

The private information of a decision maker yields a measure of power to pursue private goals that may be in conflict with corporate goals. Moreover, the many players in the internal corporate economy—shareholders, directors, managers, workers, and, sometimes, creditors—typically have at least partly divergent interests; hence, it is difficult even to impute to a firm a single overall objective. Among the objectives of the several types of players in the firm are profits, market share, growth, monetary compensation for managers and workers, quality of work, perquisites, and status. In addition, the different players may have different attitudes towards risks. Resource allocation and, thus, ultimately, organizational structure and strategy depend, at least in part, on processes of coalition formation and contest, especially when the costs and benefits of alternative allocations are difficult to measure and forecast. To go beyond the insights provided by case studies, the systematic analysis of these organizational coalitions and contests, now requires a move toward large-scale data collection among representative samples of organizations.

An active line of research has been concentrating on institutional constraints on organizations. Organizational designs are constructed and evaluated in a sociocultural context. Some designs have extensive social backing, that is, they are codified and promulgated by professional associations and schools or by government agencies. Designs also stand as markers of difficult-to-observe competencies, such as managerial acumen, and are therefore used strategically to signal such competencies. And seemingly neutral arrangements tend to become infused with moral value by members of organizations, turning means into ends. Designs may proliferate even when they make little or no contributions to productive efficiency if they serve the political or institutional purposes of subgroups within organizations or other powerful agents in the environment. Ethical and religious factors continue to play important roles as they have throughout history, such as the centuries-long effect on economic organizations of religious views about usury.

Research on organizational politics and institutional processes has made clear that organizations face strong inertial pressures. Attempts at radical redesign, especially in large, established organizations, spark political opposition and activate institutional resistance. Even without those pressures, there are bound to be transaction costs, that is, the costs of change. Opposition and costs can delay reorganizations that would take advantage of changing opportunities or enable better response to competitive threats. A core problem in explaining the spread of organizational forms is to learn how structural arrangements affect the speed and flexibility of response of large organizations.

Organizational Evolution

Although the dynamics of organizational evolution are more difficult to understand than the maintenance of existing organizational structures, a number of new developments are noteworthy. Using both theoretical and empirical techniques, researchers have developed insights into the formation of commodity and financial markets, the evolution of regulatory structure, the emergence of legal rules, the development of political institutions, and the principles of organizational change generally.

Much of the theorizing and empirical research on organizational evolution focuses on environmental factors. Some theories rest essentially on the diffusion of technological change as a driving force: when historically inherited organizations are unable to function in newer, technologically reshaped environments, new organizations arise to take their place. Some researchers have argued that the limited liability corporation evolved in direct response to the increasing pace of technological change. Other environmental factors are political in character: for example, it has been suggested that the rise of the seniority system in the House of Representatives is, in part, the result of changes in congressional constituencies that led to long incumbencies. When change is rapid, it becomes important to learn how new forms of organization are linked with processes of entrepreneurial activity. Since most entrepreneurs come from existing organizations, the dynamics of organizations undoubtedly affect the rates at which entrepreneurs are spun off and the likelihood that new organizational forms will establish footholds in competitive environments.

Some recent lines of theory and research emphasize rational-adaptive learning and copying; others emphasize competitive selection. Rational-adaptation theory accounts for both the structure and performance of organizations by focusing on the way in which large and powerful organizations respond to threats and capitalize on opportunities in their environments, which include not only the well-understood problems of availability of resources and markets but also the strategies of other organizations. Such research suggests the importance of growth by planned accretion or acquisition of new components by existing organizations, leading to the many-armed giants that seem to be a characteristic of modern times, including multicampus university systems, diversified corporations with many component subdivisions, and increasingly complex systems of military organization.

The competitive or ecological approach takes the notion of a population or system even more explicitly into account. Its central feature is that the diversity of organizations in society is seen as a kind of outpouring of new (and residue of old) organizational experiments and variations, much as the diversity of biological species must be regarded, with survival depending on the ways in which new organizations are created, the advantages they might possess in a changing environment, what kinds of environmental niches they might find, and whether previously existing organizations fail successfully to adapt to changing environmental conditions, including the arrival of new organizations.

Theory and research on organizational dynamics have developed considerable momentum. In particular, appropriate dynamic models are now in use for studying life histories of single organizations and groups of organizations. Likewise, promising starts have been made in modeling organizational learning and copying. Some convergence with other lines of social and economic dynamics have become clear, but they have not yet been exploited. Progress in understanding the issues discussed here will grow significantly as a result of building explicit bridges with work at the frontiers of dynamic modeling. In addition, this kind of approach strongly reinforces the necessity for data that uses as a sampling frame across time the universe of organizations, rather than of individuals or households (see discussion below, "New Sources of Data on Jobs and Careers"). Only with such data bases can theories of ecological process and structure be tested and new ideas based on greatly enhanced observational capabilities—rather than analogies with biological evolution or older historical theories—begin to appear.

MARKETS AND ECONOMIC SYSTEMS

Among the principal features of contemporary economic systems—markets, firms, and various forms of centralized, bureaucratic planning—markets may be the most important subtype, especially in Western industrialized countries, partly because they entail relatively little overall, expensive, organization. Markets are a fundamental type of arrangement whereby allocations and exchanges of goods and services take place. The characteristics of any market specify how individuals who enter it as agents (on their own behalf or as representatives of others) attach value to items or services to be exchanged and how they assign related costs and responsibilities. Market agents can, in principle, be anonymous: that is, the rules of the market relationship between agents do not depend upon their individual identity (except, of course, in the event of discovery of fraud or coercion), but rather on formal rules of exchange. In theory, a pure market form of organization is fully characterized by a language (or logic) of communication and choice, a set of process rules that govern communication between the agents, and a set of allocation procedures that carry agents' intentions and choices into final effect, thus clearing the market at the end of the trading period.

A trend toward markets is noticeable at the present time even in some economies strongly committed to central planning. Therefore, it is only natural that considerable current research is focused on market phenomena. At the same time, a great deal of attention is also being paid to nonmarket phenomena, especially those arising in the allocation of public goods and in handling nondirect costs (for example, pollution). The study of nonmarket phenomena is also essential to understanding the behavior and performance of most third-world, as well as socialist, economies. One central research problem in market and nonmarket economies—and in other organizational structures—is that of analyzing the interaction between incentives and information.

Prices, the sine qua non of markets, accomplish two things. First, as the common denominator of economic exchange, they make individual agents reveal information about their relative preferences, that is, the relative values, for commodities through their offers or lack of offers to buy and sell goods and services. Second, the price system aggregates all of this information so as to allocate commodities among agents. In a fully efficient market, the price mechanism facilitates a set of trades such that, when all accounts are settled, the available commodities will have been allocated so that, overall, owners are satisfied in the sense that no further trades could improve the general level of satisfaction (utility) of all agents. This is the general definition of efficient in the economic sense.

The neoclassical theory of market prices, with its normative implications for how markets should be organized in order to become fully efficient, rests on strong assumptions about the stability of agents' preferences, the ease of exchange of information among them, and their spontaneous willingness to express their true preferences in bids, offers, rejections, and acceptances. During the past decade, theoretical work and empirical findings have accumulated about the extent to which these assumptions, often grouped together under the rubric of "perfect information," can be verified, the circumstances under which they need to be relaxed or otherwise altered, and the results of doing so. A strong effort is in progress to formalize the character and results of market operations and other allocative processes involving information that is less than perfect, information that is privately held, strategically misrepresented, or intrinsically costly to obtain. This kind of research has been especially useful in analyzing policy issues related to the regulation and deregulation of major consumer and producer markets. At the same time, equally strong efforts have been made to extend the theory in a new direction, to explain financial market decisions that are predicated on inferences about future prices rather than on responses to present ones. In both of these areas, leading theoretical work has been followed up and engaged in a vigorous dialogue with empirical findings in a variety of contexts.

Public Goods and Strategic Revelation

Collective or public goods are goods whose benefits cannot be closely partitioned according to who paid how much for them—police, roads, and welfare are common examples. A very general problem is that whenever collective or public goods are to be provided through voluntary agreement, as in a marketplace or a system of taxation that includes a degree of shared decision making and voluntary compliance, public goods will be chronically undersupplied or underfinanced relative to actual individual preferences for them. There are several reasons why this is the case, but one that has received particular attention in recent years is the problem of "strategic revelation of preferences."

To illustrate strategic revelation, suppose that two adjacent property owners are both interested in building a new fence along their common property line. One owner is really willing to pay as much as $5,000 to have the fence built, and the other owner is willing to pay as much as $3,000, for a total of $8,000. Now suppose that the actual cost of construction is only $4,000. If the first neighbor would pay $2,500, and the second $1,500, each would clearly be better off with the fence than without it. But each would also be better off if they split the cost evenly—or even if their shares were reversed. So each neighbor has a monetary incentive to try to convince the other that the fence is worth less to him or her than it really is. Given the incentive to feign indifference toward a new fence, to try to "bluff thy neighbor" and get the same benefit at a lower personal cost, the fence may not be built for quite a while, or at all, even though both parties may privately prefer otherwise and even though, if each could find out the other's real valuation, they could readily reach an agreement.

As the example suggests, there is always an incentive to understate one's true demand for a collective good—whether it be a simple fence, national defense, public education, police or fire protection, environmental protection, or civic beauty—in order to reduce one's personal tax burden. At the same time, political forces often lead to considerable instability in the support of public goods due, in part, to the fact that some goods benefit one segment of society at the expense of a different, sometimes nonoverlapping, segment. The creation and elimination of water or air pollution are both examples of such an asymmetry. And support for public goods will depend on the entrepreneurial ability of political leaders to design lumps of public goods capable of attracting enough consensus.

At one time many researchers believed that no voluntary organizational arrangement or allocation mechanism could effectively solve the problem of strategic revelation or selective nondisclosure. In fact, it was proven in the 1970s that no mechanism can possibly achieve both full revelation and full allocative efficiency at the same time; in a broad class of collective choice situations, it is always in someone's interest to disguise or misrepresent his or her true preferences. However, the most extreme alternative to voluntary financing—centralized allocation—has also been shown often to lead to inefficient allocation, such as overinvestment relative to actual demand for water projects, mass transit, or other items.

Theoretical and experimental research has shown that allocation systems can be designed that can induce more efficient results in many situations. Although, in general, agents have incentives to understate their valuations of public goods, it is possible to design rules (including property rights) such that these incentives diminish or sometimes disappear. These mechanisms make possible the production of a more efficient amount of public goods (in the example above, the fence would be built; in another, a reservoir would be constructed). However, these mechanisms involve additional costs and usually still do not allow the complete voluntary financing of the public good. One such mechanism requires each participant to state the level of public service desired and the share of expenditures he or she is willing to bear. The incentive structure is so designed as to push the group toward unanimity with respect to the level of public service. Unanimity will only be attained at an efficient level of the service.

Systems of this type might conceivably be used in arriving at decisions concerning areawide services, such as airports or waste disposal arrangements, in which various communities have both common and conflicting interests.

Information Asymmetry and Transmission

The problem of strategic revelation and, more generally, of privately held or asymmetrically distributed information, extends to markets for private goods as well. Consider the following recent, but already classic, analysis. Suppose that each potential seller in a dispersed market for used cars knows from experience the value or quality of his or her own car (which may be thought of as an index of how long that car is likely to continue running reliably) and that this quality cannot be directly observed by nonowners. Buyers thus do not have direct information about the quality of individual cars, though they can acquire statistical information (say, from a consumer organization) about the overall proportion of "lemons" on the market. If any high-quality cars are offered, buyers would be willing, based on the statistical odds, to pay a price for a used car that is higher than the value of the lowest quality car, but lower than the price that potential sellers of highest quality cars would consider acceptable. At this price, owners of most used cars will keep their cars, but owners of lemons will be happy to sell theirs. In this analysis, only lemons will ever be put up for sale voluntarily, purchasers who might prefer to buy higher quality used cars will find none for sale, and the market prices for used cars will end up being equal to the value of lemons.

This analysis of how asymmetrical information can shape a market explains why used car buyers look for owners who "must sell"; why people pay so much attention to whether a used car is "clean," which has the virtue of giving visible evidence in a circumstance in which the most important information is hidden; and why used car dealers, whatever their business ethics, have so much trouble maintaining or establishing a respectable reputation. Moreover, this analysis shows that the used car market (the "market for lemons") is necessarily an inefficient one, since asymmetry of information leaves unexploited gains (mutually worthwhile deals) still unmade after trading is completed.

To be sure, the analysis assumes an extreme case, because in reality buyers can gain some information about the nature of a particular car by taking it to a mechanic for testing, and sellers operate under social constraints such as product liability and warranty requirements. Each of these factors affects the structure of information and prices in the market. Yet such remedies for imperfect information involve real economic costs, which are called transaction costs. Such costs have been the subject of intense efforts at measurement. Insofar as they arise out of the basic conditions of the market for "lemons," transaction costs redistribute or simply amortize to some degree the inefficiency that is due to informational constraints.

These several cases illuminate a general principle about the effects of private or imperfect information: it is costly to extract information about the characteristics of individual economic agents or commodities, and the relative costs of gathering and monitoring such information strongly affect the performance and efficiency of the market.

The extension of this work on market information and, especially, its fuller integration into neoclassical price theory and principles of market design, is an important challenge to theoreticians and experimenters; it may also prove of future value to agents, managers, and regulators.

In this context, an outstanding difference between market mechanisms and central planning is decentralization versus centralization of information. Markets require relatively limited transmission of information about preferences and costs because prices are extremely efficient in encoding the information that is relevant to transactions. In fact, it has been proved that, in a steady-state economy, no informationally decentralized mechanism which has any less "transmission capacity" than the price mechanism can assure efficient resource allocation.

But for individuals to make good decisions based on such decentralized information is not always easy. An informational aspect of economic mechanisms that has been studied recently is their computational complexity. As one example, the complexity of linear programming problems and algorithms has received much attention, and a beginning has been made in measuring the complexity of competitive markets. In its more abstract aspects the study of complexity is a bridge connecting current research in economics (for example, problems of decentralized resource allocation) with research in information and computer sciences (for example, problems of distributed computing), and certain branches of pure mathematics.

Regulation and Deregulation

During the past 20 years, largely fueled by theoretical developments of the sort discussed above, the study of regulation, deregulation, and the consequences for market performance of variations in the governance of commercial exchange has accelerated. Traditional regulatory efforts to maximize the public welfare are based on shifting the incentives or extracting private information that otherwise might lead firms to maximize profits at the relative expense of environmental or consumer health and safety. The question is to what degree those regulatory solutions are flawed because they distort the incentives of firms to minimize production costs, inviting costly efforts to circumvent regulation or requiring costly monitoring to ensure compliance. In a variety of industries—airlines, telecommunications, trucking, broadcasting—the view that public regulation is necessarily or probably the most effective remedy for inefficiencies or monopolistic domination and manipulation of prices and output has been particularly questioned. Many of the questions have arisen from studies on how the costs and incentives of firms are affected by a variety of alternative mechanisms for disciplining them, including quasicompetitive mechanisms and the availability of legal recourse to their customers.

Regulation is now known to be an inherently complicated political and legal process, influenced not only by Congress, the President, the courts, and the regulatory agencies, but also by the industries subject to regulation, their competitors, and consumers. Private interests often spend considerable resources in efforts to deflect the impact of regulation by trying to influence the regulatory agencies or the political officials who supervise them.

An example of how research advances can inform regulatory policy is found in the case of cable television. The main issue of the 1970s was what effect cable television would have on the viability of "free" over-the-air broadcasters, especially independent stations using UHF channels. The underlying cost structure of cable television, the demand for subscriptions to it, and the viewing patterns of subscribers were analyzed. On the basis of these studies, the winners and losers under various scenarios for cable development were determined. The research indicated that cable television would primarily increase the likelihood of effective competition with the three national networks, in part by improving the market position of UHF independent stations. These results influenced subsequent decisions by the Federal Communications Commission and by the several congressional oversight committees to relax regulatory constraints on cable television.

Similar kinds of research on costs, demand, and competition in the airline industry had a bearing on policy questions concerning the potential effects of deregulation on fares, safety, quality, and the availability of service to smaller communities. This research played an important role in convincing Congress that airline deregulation would not lead to a substantial reduction in service, but would lead to substantially lower fares, except for some smaller cities.

Both of the preceding examples referred to the use of research in making decisions to regulate or deregulate. Research has also played a significant role in addressing questions related to the consequences of changes in regulatory methods. For example, rapidly rising energy costs during the 1970s caused most electric power executives and regulatory officials to examine new methods for pricing electricity. The central issue was whether and how to introduce "peak-load" pricing: prices that vary over time to reflect differences in cost and demand conditions. In order to estimate the consequences of alternative pricing methods, the federal government financed several field experiments with peak-load pricing, to find out which form of pricing produced the greatest efficiency gains (prices could vary literally from moment to moment at one extreme or could change only seasonally at the other), and for which classes of customers a switch of pricing methods would yield a gain that exceeded the implementation costs.

Rational Expectations

A very active area of contemporary research deals with market decisions about the movement of capital. One common feature of such financial transactions is the inherent lack of definitive information regarding the future payoffs or values of the assets changing hands. Theories of asset pricing and risk sharing have depended greatly on identifying the expectations about future events that are held by participants in financial markets and have usually assumed that future expectations were largely based on—or biased by—the trend of past events.

An important innovation of the 1970s was a set of theories about "rational expectations." These theories hold that individuals form expectations about the future that are based on all of the currently available information and that agents in financial markets can be presumed to use the best techniques available for drawing prospective inferences from these data. Rational expectations have the property of being "optimal" (that is, the best possible statistical forecasts) on the average. Stated most strongly, the hypothesis of rational expectations implies that financial markets are fully efficient, with the asset prices prevailing at time taking into account all available information on possible price movements. The innovative feature of such theories is that past price movements are taken to have no particular value in predicting future prices. The only way to "beat the market" consistently under these assumptions is to secure inside information. When applied to financial markets and to those nonfinancial markets that affect the economy as a whole, these efficient-market hypotheses lead to the further conclusion that taking planned government action to counteract the business cycle will prove ineffective because by the time such action is undertaken, it will have been anticipated and appropriately discounted or adjusted to by the market. If correct, which is a matter of ongoing research and debate, such a conclusion would have major implications for government economic policy.

The early testing of rational-expectations theories on actual market behavior has not supported those theories in their strongest form. For example, the stock market fluctuates excessively in relation to the underlying determinants of stock value: that is, the market is more volatile than efficient-market hypotheses permit. Related research on the maturity structure of financial markets—the distribution of times when credit instruments bearing various interest rates come due—has also rejected the strong rational-expectations hypothesis. This kind of empirical testing has been made possible by improved models of asset pricing with testable empirical implications, development of econometric methods of dealing with rational-expectations hypotheses, and the availability of excellent data on asset returns.

The failures of a strong version of rational-expectations theories at the aggregate level of stock price movements have led to intensive examination of their underlying statistical assumptions and to their revision. With respect to the stock market, innovations have included new theories of rational-expectations equilibria, which can produce random cyclical price movements without an external cause, and particularly of "speculative bubbles." A speculative bubble exists when the market price of an asset diverges significantly from the price that would be indicated by those fundamental factors that should determine its value. But in addition to its basic value, a stock's price is determined by the possibility of reselling it to someone else who, in turn, may expect to resell it again at an even higher price. This speculative process has been shown to be theoretically consistent with rational-expectations equilibria in which the probability of asset-value growth is high in comparison with the probability that the bubble will burst. Bubbles can develop in any economy in which the interest rate is less than the growth rate. In such economies bubbles may actually improve the allocation of resources—which is the underlying issue, not whether smart people can make easy money on the stock market.

The rational-expectations approach has been applied in areas other than financial markets. For example, the life-cycle and permanent-income theories of consumption and savings behavior imply that a person's consumption rate is determined by income and wealth considered over the lifetime. A person's current level of consumption thus depends not only on present income, but also on the person's expectations concerning the present value of future income. Assuming that these expectations are rational, the life-cycle hypothesis implies that since decisions are based on overall expectations, levels of consumption in one time period have no direct impact on the consumption in the next time period. Empirical studies at both national and household levels are largely consistent with this hypothesis of independent variations through time, but they also show that consumption is nevertheless more closely related to fluctuations in current income than the rational-expectations, life-cycle assumptions imply. This finding has shifted theoretical and empirical investigation to the effects on consumption of constraints that make it more or less difficult (or impossible) to borrow against future income.

An important related issue is the effect of government budget deficits on real interest rates. The rational-expectations hypothesis implies that taxpayers know that the government will have to impose taxes in the future to retire the debt. Since the present value of future taxes is equal to the amount of the debt incurred, it follows that government borrowing is equal to future taxation. Thus, according to the hypothesis, explicit taxation and deficit financing should have exactly the same effect on savings and real interest rates. However, direct empirical tests of this theory have proven inconclusive. If further research using better data on private consumption and taxpayer behavior can establish that real interest rates are affected by the method the government uses to finance its spending, a major empirical issue in present debates over government deficits will be resolved.

The rational-expectations approach has been theoretically rigorous and innovative, has generated controversy, and has stimulated systematic empirical research. It has not survived fully intact; a number of empirical tests have produced results that either run counter to its boldest hypotheses or are inconclusive. This outcome poses both strategic and methodological challenges. The strategic challenge is to abandon an "either-or" approach to rational expectations and to press theory and empirical research forward by identifying the conditions under which rational expectations hold sway and the conditions under which other forces impinge or dominate. The methodological challenge is to make more concerted efforts to measure directly how expectations are actually formed by individuals in empirical settings. Research up to now has been largely based on market outcomes from which plausible expectations of actors are inferred. A more direct approach could be effected by using combinations of aggregate data analysis, experimental simulations, and panel studies to learn directly how individual agents form expectations and use them in decisions on financial holdings, savings, and borrowing.

CONTRACTS

In theoretical terms, contracting relationships become important where classical market assumptions or requirements fail or are infeasible—as, for example, where the products or services to be bought or exchanged are not well defined, the risk of precommitment for uncertain exchanges is too great, transactions are nonrecurring, inputs are highly specific, or collective goods are at stake that must be shared over a long term. In short, conditions of imperfect or asymmetric information often lead to making contracts. The study of contractual relationships has become a major research front, with a diversity of methods and theoretical perspectives. It has proved to be an important area concerning the interplay of individual preferences with organizational structures, an area comprised principally of analyses about information flows and the information underlying decisions.

Principal-Agent Models

One main line of work on contractual relationships is the study of the principal-agent contract. In many situations, individuals contract for goods or services that cannot be described fully prior to delivery or performance—for example, a landowner seeking the services of someone to farm his or her land, or a client seeking the services of a lawyer, doctor, or financial manager. In these cases the individual who seeks the service, called the principal, does not ordinarily find it worthwhile or feasible to specify in advance exactly what constitutes adequate performance. This contractual vagueness provides opportunities for the provider of the service, the agent, to gain an advantage under some circumstances. At the heart of principal-agent research is the careful modeling of precisely what information each of the parties has at the outset and what information each is able to obtain during operations, which is particularly important when asymmetry of information exists about some aspect of the environment in which the contract is operating.

One of the first applications of principal-agent modeling was to sharecropper contracts. This type of arrangement between landowners and tenant farmers, commonly observed in both the United States and other countries, commits the parties to share the agricultural output in fixed proportions. This pattern was difficult to explain using previous generally accepted theories of economic organization, because such a contract results in the landowner and the tenant farmer bearing the same amount of risk in regard to their total compensation from the activity. Neoclassical utility theory predicted that the risk should be borne differently, with the landowner bearing more risk than the tenant, since a bad crop (and, hence, a small share to each party) would be much more hazardous to the relatively low-income tenant farmer than to the landowner.

Under this theory, the only explanation for the observed pattern of equally shared risks seemed to be that sharecropping is not a contract voluntarily entered into between equals. In this country it was most notoriously an arrangement that became common after the Civil War between white southern landowners and poor, illiterate former slaves in a social setting that terrorized and oppressed black farm families. But the system has appeared in much less coercive settings as well, which suggested that an economic rationale aside from exploitative subjugation might also be at work. Principal-agent considerations identified such a rationale. Consider the information-monitoring difficulties that would arise in a contract in which landowners bore most of the risk. Also consider that a contract with unequal risk would give tenant farmers a return that does not vary in proportion to the variation in total output, and so there would be incentive losses in the sense that tenant farmers would not expect their return to depend greatly on their efforts. A tenant farmer obviously knows the amount of effort exerted, but a landowner typically can observe in detail the level of activity of the farmer only at rather high cost. Having modeled carefully the information-monitoring difficulties in this problem, researchers then asked what kind of contract would lead to the largest net output (total output minus the cost and effort of producing it). It has been shown that the kinds of sharecropper contracts and related organizational structures observed were quite consistent with the model's predictions.

The sharecropper problem is an example in which the principal-agent paradigm is helpful in understanding why a particular type of contract evolved. The analysis discloses that, depending on the assets available to the two parties, the possibilities of close monitoring, and other specifiable factors, sharecropping may be a very efficient organizational form relative to the alternatives (such as land rents or wage labor) developed across centuries of agriculture. There are many problems of organization for which there simply do not exist centuries of experience, and the careful analysis of such problems, using the principal-agent paradigm, allows one to identify, at least in crude form, what kinds of contractual structures may be optimal. For example, many questions of public policy concern kinds of contracts and transactions firms should be allowed to engage in, and these problems can only be analyzed with some theory in mind as to why firms want to use particular contractual arrangements. A specific example, addressed in recent years, is retail price maintenance, in which a firm sells goods wholesale to a retailer but controls the price at which they can be resold to the retail customer. This practice has been both explicitly legal and illegal at various times and places. The discussion of whether it is in the customer's interest that such practices be prohibited cannot be answered without some explanation as to why the original firm would want to control the retail price. Consistent explanations were lacking until the advent of principal-agent models, which define the business incentives and facilitate careful consideration of the conditions under which such practices benefit or cost the ultimate consumer.

In a similar matter, the question of whether it is in the general interest to encourage or discourage mergers between large firms requires an understanding of the purpose that mergers might serve. In the case of a vertical combination (that is, a merger between a firm and one of its suppliers), it has long been a question of what is the advantage of such a merger over a sophisticated contract between the two firms. A focus on the difficulty of monitoring contracts between separate firms, in contrast to internal processes in a single (merged) one, has led to a better understanding of the purposes of vertical combinations. This understanding can provide the groundwork for more informed policy regarding such business practices.

One striking success of the principal-agent is in managerial accounting, which deals primarily with how firms use accounting information for internal operations, management, and compensation (as opposed to financial accounting, which has to do with the use of a firm's accounting data by people outside the firm). The principal-agent model has revolutionized the way managerial accounting is taught in business schools; to a large extent, the principal-agent model has become managerial accounting.

The opportunities for further extension and application of the principal-agent model are considerable. It has begun to be used extensively in international trade problems to analyze relationships between industries in various countries and the governments that must set rules for international competition with only partial information as to the internal structure of the industries. There have been initial attempts to study the evolution of organizations over time and to understand what forms of principal-agent contracts best allow organizations to have the flexibility to respond to changes in their environments. And many of the techniques of the principal-agent model are applicable to bargaining and arbitration problems in which an arbitrator is only partially aware of the costs and benefits of various alternatives.

Bargaining, Negotiation, and Repeated Interaction

A bargaining situation is one in which agents make offers to one another and accept or reject them in the context of agreed-upon organizational rules and understandings. Some of the variables involved in bargaining are the level of agents' uncertainty about the value of an offer, how impatient they are to come to an agreement, and their personal attributes or preferences. Much of the work in this area relates to information problems in markets, discussed above.

The broad outlines of a theory of bargaining are becoming clear. For example, game theory—the mathematical analysis of interdependent, partially conflicted situations involving two or more decision agents—indicates that if all information is known to all parties, the first offer made will always be acceptable and accepted, and any trade that is efficient will be made. If all agents know the value of all offers, there is no point in bluffing or holding out. In this full-information case, the consequences of different kinds of rules for making offers can be easily calculated, which sets of rules give which players an advantage can be determined, and players' preferences for different sets of rules can then be predicted.

In contrast, when there is private or proprietary information, bargaining, takes longer because the agents' efforts to elicit information about one another become part of the action. In this case, bluffing and holding out—or believing that the other player is doing so—may be a useful strategy, but it may also prevent the completion of efficient transactions. Recent theoretical results yield models of sequences of offers in private-information situations that make it possible to estimate the value to agents of alternative rules. It has also been possible to account mathematically for the fact that when a bargaining situation occurs repeatedly, bargainers may develop reputations for toughness that affect outcomes.

Applications of this new research to breakdowns in bargaining relationships, such as in litigation, wars, and strike arbitration, are tentative but promising. Compulsory arbitration, for example, generally tends to reduce the cost of failing to reach a bilateral agreement and may have a chilling effect on bargaining. Final-offer arbitration schemes—ones in which, as in professional baseball disputes, the arbitrator is required to choose the offer of one side or the other rather than produce a compromise—create a powerful incentive for both sides to produce realistic offers. Further insights into these systems can be generated by the study of histories of union negotiations from the standpoint of contract theory, bargaining theory, and alternative theories of behavior in structured, repetitive, situations.

While the early formulations of principal-agent theory dealt mainly with static situations, theorists have recently been able to take dynamics explicitly into account, that is, to model how an individual, when deciding how to act at a given moment, anticipates future interactions as well as immediate payoffs. For example, even if an individual has a short-term incentive to cheat on a contract, he or she may choose not to do so because the long-term losses from cheating outweigh the short-term gains. Whether this is so depends in turn on how the cheated party reacts to cheating. These kinds of considerations have led to the formulation of a theory of reciprocity (that is, contracting in cases of repeated interaction).

One key to this theory centers on the capacity of contracting parties to detect and punish cheaters. If cheating is easy to discover and is regularly punished, the incentive to cheat will lessen; occasional repetition of the punishment may secure high levels of compliance. But whether or not cheating is punished depends on whether contracting parties are committed to carrying out punishment even if it turns out that punishing the violator is very costly. The theory has important implications for such problems as nuclear deterrence. If a nuclear power were to use a few nuclear weapons to obtain some specific foreign policy goal while retaining the capacity to initiate a general nuclear war, other powers would find the initiation of punishment very costly, since it would involve the risk of beginning a general war with consequent massive nuclear destruction. No matter what the powers promise themselves and others in advance, a potential cheater still can reasonably wonder whether nuclear threats would actually be carried out. Nevertheless, simulations indicate that even if the chance that swift and severe punishment might be carried out is relatively small, that chance is likely to sustain deterrence.

The notion of reputation or credibility is also central to the theory of repeated interactions. People in such interactions are sensitive not only to the future and their understanding of how long the relationship may continue, but also to the existence of private information. For example, in an international disarmament negotiation, both parties usually feign a large degree of indifference as to whether or not an agreement is reached. But this gains little advantage in the negotiations, unless the other party believes the apparent indifference to be real—which it ordinarily will not. Not to express indifference when one has a reputation for doing so, however, is highly credible.

Further progress on theories of repeated interaction and bargaining and negotiating sequences will require developments in several techniques. The mathematical skills required for their fullest development are not yet widespread, and the underlying extensions of game and bargaining theories are only beginning to be well understood. Moreover, many techniques of field observation and statistical analysis appropriate for work in this area are not adequately developed. Finally, laboratory methods that promise to be useful for research in this area have to be refined. In particular, the training of researchers in empirical research design, mathematical methods, and statistical techniques requisite for experimentation has to be expanded and links to researchers in cognition, language, and artificial intelligence have to be strengthened. A major interesting avenue of future research lies in the application of computer technology, which offers the opportunity of more thorough interactive experiments. The development of standardized software and operating systems would decrease the costs of performing such experiments and generate a large body of empirical data free from the effects of uncontrollably differing experimental designs.

JOBS, WAGES, AND CAREERS

The study of jobs and labor markets incorporates virtually all the issues that arise with respect to individual and collective choice: imperfect information, varying incentives and expectations, problems of negotiating and enforcing contractual agreements, and pressures of organizational inertia and change. Theoretical and empirical study has focused on the effects of the business cycle, the protocols used by individuals in preparing and searching for jobs, the practices of employers in recruiting workers, implicit understandings between employers and workers, demographic differences in earnings and job assignments, migration, and changes in the content of work. When viewed closely, employment contracts turn out to be an idiosyncratic and imperfect match between workers' abilities and employers' needs. Complex systems of formal and informal labor-management and labor-capital relations develop within organizations as ways of trying to ensure that future actions and events will be consonant with earlier agreements and decisions. This kind of study has been cross-pollinating with studies of other kinds of long-term agreements whose implementation is a matter of subsequent bargaining, monitoring, and exchange, heavily influenced by wider political circumstances, such as the contracts that electrical utilities make for fuel deliveries or that owners and tenants make for rental housing.

The depth and breadth of research relevant to jobs, wages, and careers and the extent of public interest in this set of topics lends the area a perennial interdisciplinary vitality. At the same time, a convergence of research interest and political imperative suggests the usefulness of a sustained effort to generate substantially better and more accessible collections of observational data on these topics.

Unemployment

Few topics in the behavioral and social sciences have attracted as much interest as unemployment. The subject is complex, not least because the term covers a number of very different ways in which unemployment may come about, and a distinct line of research corresponds to each of them.

First is cyclical unemployment, in which jobs disappear and reappear due to overall swings in the current or anticipated profitability of production, in short, due to the business cycle. Second is frictional unemployment, in which workers, after either voluntarily quitting or being fired from a job, seek different jobs and therefore are unemployed while the job search takes place. Third is structural unemployment, in which jobs disappear or do not exist because of such features as declines in the demand for particular kinds of labor, sometimes resulting from technological change; gaps between the skill demands of jobs and the skills of available workers; employers' negative beliefs about some workers' capacities for particular jobs or some forms of prejudice, which tend to produce an unemployable underclass; and other factors that create a long-term gap between the demand for and supply of particular workers. The degree to which cyclical, frictional, or structural factors predominate in a given level of unemployment has important implications for the type and effectiveness of programs or policies that would be appropriate for reducing unemployment.

Recent research with aggregate data on the business cycle and its temporal covariates indicates that overall wage rates respond only slowly to swings in productivity associated with downturns in the business cycle. Most of the adjustment to such downturns takes the form of layoffs and reductions in hours worked rather than lowering of wages. Some studies of contract structures for employment, marketing, sales, and delivery, especially focused on the lags and staggering of price adjustments, suggest that the reason output and employment often are highly sensitive to unanticipated shifts in demand is what is called price inertia—that prices, like wages, respond only slowly to demand shifts, possibly because of the costs associated with making changes in prices. This hypothesis certainly is consistent with the fact that both overall money prices and wage levels are relatively "sticky" and unemployment is highly sensitive to the business cycle. However, as yet there is no theoretical explanation for price and wage inertia. Clearly, arriving at such an explanation should have high priority.

Better data on contract terms—both explicit and implicit—and the details of price setting by firms and industries is essential for increased understanding of the dynamics of wage adjustments. Such data would also increase the ability to assess the effectiveness of economy-wide strategies for damping swings in unemployment and planning compensation schemes and layoff policies. Theoretical model-building is necessary in order to find ways to capture and test underlying ideas about how business cycles work. This kind of knowledge is also critical in understanding the degree to which both employers and workers profit and lose from unanticipated swings in the economy, and how these matters affect the commitment and well-being of managers and workers. Knowledge about these labor-market dynamics may help both employers and employees develop principles to guard against disastrous losses in difficult times and yet provide adequate incentives for performance.

A separate realm of research focuses on frictional unemployment, that due to between-job searches. Survey data of manufacturing jobs and of households reveal that average spells of unemployment are short, with the highest turnover rates found among young and secondary workers within families. This finding tends to support the view that frictional unemployment is rather common. Policies aimed at improving the efficiency of search procedures and job information could reduce this kind of unemployment, although such positive effects might also increase the attractiveness of undertaking searches for better job matches, thereby increasing frictional unemployment.

The same surveys show that, despite the shortness of the average period of unemployment, most of the total days of unemployment are accounted for by a minority of individuals who are out of work for prolonged periods each year: this finding indicates that some structural factors are responsible. Although there has been little empirical work on structural unemployment, a striking finding of theoretical studies on job searching and recruiting is that relatively minor mistakes by individuals in assessing the job market can be expected to result in excessively long periods of unemployment for some workers.

Research on unemployment has been limited by the scarcity of detailed and reliable data on search protocols and decision behavior. Additional data on these topics from the United States and other industrial economies (see Chapter 4) could be used to investigate incentives associated with income-security programs. Such data would make it possible to assess how effective—if at all—providing better job-information mechanisms or building different incentives into unemployment insurance programs or labor contracts might be in reducing unemployment.

Implicit Labor Contracts

The search for good matches between workers' abilities and firms' needs is costly for both employers and potential employees. It would be very costly—if not impossible—for employers to try to improve the quality of such matches by supervising workers in every detail of their jobs or attempting to measure their productivity more precisely. As an alternative method, employment contracts explicitly or, more often, implicitly provide incentives for workers to be loyal, diligent, and to acquire the specific job-related skills that can improve their performance. However, these contracts are imperfect, and monitoring and enforcing them can be costly. Examination of the incentive and information properties of such contracts as well as their relation to the legal system and the actual behavior of employers and employees shows that they often have unintended consequences, some desirable and some not.

Some significant findings have emerged from the analysis of long-term employment histories that have become available from research projects that were initiated in the 1960s. In contrast to the stereotype of "Americans on the move," it has been found that following an initial period of considerable job mobility in their 20s, most workers settle into stable employment patterns and change jobs infrequently after age 30. (There is far more job mobility in the United States than in European countries and Japan.) Other data show that the ratio of earned wages to measured productivity rises with age. One possible explanation for this steady age-related growth in wages is that workers acquire skills on the job that are not measured by productivity data, and apparent wage premiums for simple seniority actually reflect payment for these unmeasured, experiential skills. An alternative explanation is that high wages for older workers hold out promises of future (delayed) compensation to younger workers in return for staying with the firm, thereby strongly discouraging younger workers from looking for jobs elsewhere, which would require employers to engage in costly searches for replacements.

Neither of these proposed explanations has been definitively confirmed or rejected, and the truth may turn out to be some mixture of the two. But the alternative theories have very different implications for such policies as anti-layoff legislation, minimum wage laws, and vesting of pensions. For example, neoclassical economic theory suggests that a minimum wage will exclude from short-term employment markets those workers—particularly young ones—whose productivity is below that minimum. The resulting reduction in the competitive pool of workers would increase the wages of somewhat more skilled workers, who would therefore support minimum wage legislation. But more significantly, in labor markets that provide long-term incentives for workers' diligence and learning (in the form of delayed compensation such as premiums for seniority), a minimum wage may serve to discourage long-term employment and related skill development among just those workers whose immediate and potential long-term productivity are both above the minimum—because the higher wages that need to be offered at the outset must be financed by reducing the delayed wage premium for long-term employment. The resulting wage profile from the minimum on up may not increase steeply enough with seniority to keep many people from moving around among employers, preventing them from learning the unique job-specific skills that permit highest productivity.

Similar issues of implicit contracts involved compensation also affect the question of how to stimulate corporate managers to achieve high levels of productivity. Research using the theory of implicit contracts and incentives has shed some light on why the very highest corporate executives receive much higher salaries—two and three times as much—than those not far below them in the corporate hierarchy. These differences are not consistent with explana-

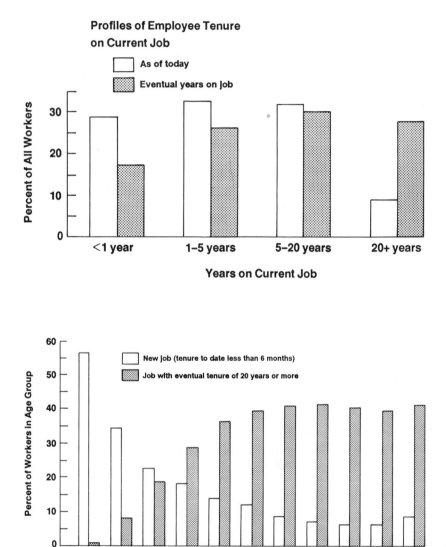

Profiles of Employee Tenure on Current Job

tions that match compensation with current or even future expected productivity, but can be understood as delayed compensation for past productivity.

Although there are some existing data that can illuminate competing theories of job markets and the productivity of workers and executives, knowledge would be greatly improved by longitudinal data on employee behavior as a function of the initial contract of employment and related benefits, customs, and expectations in the employer-employee relationship.

LONG-TERM EMPLOYMENT   How long do American workers stay with the same employer? Are they more restless, or treated by employers as more dispensable, than workers in other industrial countries? In a time of rapid technological and social change, many workplace and labor policies, such as programs of on-the-job training and retraining, depend importantly on assumptions about the nature of work careers.

The labor market in America is often viewed as turbulent, with frequent shifts from one employer to another, especially in contrast to European traditions of job stability or the *nenko* system of lifetime employment in Japan. National survey data on job tenure certainly suggest, on their face, that lengthy employment is atypical. For example, the median tenure among all American workers in their current job is only about 3½ years; at any one time, 40 percent have been in the same job for less than 2 years and no more than 1 in 10 have been in the same job for 20 years or more.

However, sophisticated analyses of survey data yield a rather different profile of job tenure. This figure shows the data carefully adjusted for years in the labor force and projected through time. Although many workers are new to their current jobs, a large fraction will stay on them for many more years. From data based on national labor force experience since World War II, nearly 6 in 10 workers surveyed in 1978 were in a job that would eventually last at least 5 years; three of ten could expect it to last 20 years or more. Of course, age matters; young workers are likely to be in new jobs, and their current jobs are unlikely to be long-lasting. The tenure profile with age is striking: as workers age beyond their teens and early twenties, there is a steady decrease in new job starts and a corresponding increase in long tenures. But then these trends level off: after age 40, patterns of tenure and turnover show virtually no further changes.

Furthermore, black and white workers differ little in the likelihood that their jobs will be of long tenure. Job discrimination is evident in other data: blacks work in lower paying jobs on the average, and a substantially higher proportion have no job. For women, job tenure averages about half that of men, and only 15 percent of women (compared with 37 percent of men) will be in their current job for 20 years or longer. Overall, men and women average about the same number of jobs (10½) in their careers, but women spend much longer periods out of the labor force.

Job Segregation and the Gender Wage Gap

The social issue of job discrimination by race or gender is of long-standing interest. Persistent gaps in wages and job status are found between white males and other demographic groups. According to the neoclassical economic theory of this labor-market phenomenon, these gaps reflect intragroup differences in average individual experience and investment in human capital (education and training). According to the theory, firms have little or no incentive to discriminate for noneconomic reasons because they would be punished by economic competition and lose profits if they did so. Research on exchange under asymmetric information provides a different perspective of the observed phenomena. If firms must forecast the productivity of potential employees from readily observable characteristics, then any historical differences in average productivity due to different levels of investment in human capital by race or gender, or to negative effects of irrational prejudice, or to active hostility in the workplace may lead firms to use criteria of race or gender for screening and job assignment. The motivation for such discriminatory behavior, even if it is illegal, may be economically rational from the firm's point of view in the absence of specific information on the skills of the individuals who face such discrimination—unless, of course, the dangers of expensive litigation are sufficiently large. Consequently, the incentives for hiring or promoting disadvantaged people may well be slight, thereby reducing even further the opportunities for people from disadvantaged groups to improve their skills. Such a pattern results in a self-fulfilling prophecy, self-perpetuating discrimination.

In recent years, research on discriminatory processes concerning where people work and how much they are paid has advanced more rapidly with respect to gender differences than racial or ethnic differences. The gender distribution of the paid labor force in the United States has shifted substantially in the last century. During the years of heavy industrialization (1880–1940) men constituted more than four-fifths of the total labor force, and only one-tenth of all married women worked outside the home. With the rapid expansion of service and white-collar employment after 1945, the proportion of women employed increased steadily. Women now constitute 43 percent of the paid labor force, and more than 50 percent of all married women are employed. Moreover, the character of women's working careers is becoming more like men's in two major ways: women are working many more years than before in full-time rather than part-time jobs, and they are working uninterruptedly, in spite of childbearing.

Despite these changes in women's participation in the paid labor force, their average earnings have remained remarkably stable at about 60 percent of men's for full-time, year-round workers. This wage gap is due largely to the segregation of labor by sex: that is, men tend to be concentrated in occupations that receive higher wages, and, within occupations, at jobs that are at the higher end of the wage scale for a particular occupation. Like the wage gap, sex segregation across occupations has been remarkably persistent. Its level in the United States did not change much from the turn of the century until the 1970s, even though particular occupations came and went and others changed radically in composition from predominantly male to predominantly female or vice versa. Of 503 occupational categories in the 1980 U.S. census, 275 exhibited a sex ratio of four to one or more, and about half of all workers were employed in such relatively segregated jobs.

Accounting for wage differences and job segregation in ways that withstand the test of empirical evidence has proved a recalcitrant problem. One theory that tries to account for the excessive number of women in lower paying occupations on the basis of personal characteristics and rational choice: in anticipation that they will interrupt their employment to bring up a family, women make different occupational choices from men, selecting occupations that qualify them for jobs they can leave and reenter easily. Since men do not interrupt employment to bring up a family, these "easy-come, easy-go" jobs would be highly segregated, and since experience premiums would not be sought, the jobs would be at low wages. However, research has seriously challenged this trait-based theory. For example, empirical studies show no greater tendency for women whose working careers were interrupted rather than continuous to be concentrated in heavily female-dominated occupations. In addition, the negative effect on earnings of time out of the labor force appears to be no different for male-dominated occupations than for other occupations. In addition, the rates at which the earnings of women in predominantly female occupations increase with experience do not differ from those for women in less segregated occupations. In short, women in male-dominated occupations earn more than women of measurably equivalent talent and training in female-dominated ones and experience no greater tendency to be "punished" for family interruptions. So for women freely to choose careers in "women's work" does not seem an economically rational choice, whatever their family plans.

An alternative theory about why women enter segregated occupations, also based on personal characteristics, is that sex role socialization leads women strongly to prefer certain occupations for reasons unrelated to economic rationality and therefore to choose the training and education appropriate for them. However, the research evidence has also accumulated to the contrary for this theory. Young men and women both display considerable movement within the labor force, including a moderate amount of mobility across sex-typed occupations. Women have, in fact, proven quite responsive to newly available opportunities at increased wage rates for at least 100 years, including: the dramatic increase of women clerical workers relative to men from 1880 to 1900; the flood of women into previously male jobs during World War II; the rapid movement of black women out of domestic service and into clerical work after 1950; and the more recent sharp increases in the proportions and numbers of women becoming doctors, lawyers, and coal miners. The evidence is that women's aspirations are shaped by their expectations about what kinds of occupations are accessible to them, rather than by fixed preferences for particular jobs.

Since the actual preferences of women have been shown to play a limited role in explaining their segregation into lower paying occupations, research attention has turned to the ways in which other influences affect the assignment of women to "women's work." For example, one line of accumulating research has investigated the role of people who make available information about various occupations, especially about their entrance requirements; the research shows that such information is presented differently to male and female students, from preschool to vocational training programs. Other research focuses on the detailed behavior of employers who characteristically do not hire women into men's jobs and vice versa, or who steer prospective applicants into gender-typed openings, and of unions, particularly those that have histories of excluding women or of not representing demands for pregnancy leave and other benefits that favor women, while supporting seniority benefits and other demands that favor men.

Analytic attention has also focused on the effects of husbands discouraging their wives from job training or employment that would modify their regular home activities or of persuading their wives to leave their jobs when the husbands relocate. These effects may be reinforced by differential job ladders and training programs for men's and women's positions in firms, firm-wide job evaluation systems that underestimate the training and conditions of jobs held primarily by women, and institutionalized requirements in many jobs held primarily by men to work overtime or relocate at the employer's bidding. The wage gap may itself perpetuate a pattern of household decision making in which a husband's occupational opportunities and choices come first because they are more critical for household income. The lower valuation accorded to work by women by the structure of wage rates becomes, in effect, a self-fulfilling prophecy.

What is especially promising in many of these new lines of research is the focus on developing longitudinal or process data, which permit researchers to discriminate between competing theories that may all be inferentially consistent with data on outcomes alone.

Technology, Migration, and Mobility

The organization of work and the evolution of working careers has long been an active and productive research area. One area of current controversy concerns how present trends of technological change affect the quality of workplaces and career opportunities. Do these changes increase or diminish skills, responsibilities, and commensurate rewards on the job? A second area of controversy surrounds the factors that shape the organization of work. One traditional view is that, when technologies change, firms adopt the work arrangements that best achieve administrative and technical efficiency, generally subject to a degree of bureaucratic lag. However, recent studies of technological change document considerable discretion in how specific jobs are organized, indicating that imperatives other than technical efficiency may systematically shape work redesign. Some researchers suggest that detailed distinctions among jobs serve to reduce the importance of high skill levels, diminishing workers' control of job activities and dividing various subgroups of the labor force. Moreover, there is evidence that powerful people in organizations often redefine work roles to suit their idiosyncratic interests and abilities, not necessarily those of the firm.

Yet another controversial area of research involves the link between schooling and career. One view is that employers define jobs in terms of productivity requirements and then use educational credentials to decide whether individuals are likely to satisfy those requirements. An opposing view is causally the opposite, namely, that organizations define jobs and career paths around their workers' educational attainments. Research has yet to provide evidence for selecting between these two views.

Recent analyses of labor markets focus attention both on career movement within organizations, including the manner in which vacancies occur at the top of an organizational ladder and move down them as a chain of promotions, and on movement among firms. Different determinants and consequences are associated with these two types of career mobility. For example, women, racial and ethnic minorities, the young, and the old are thought by some to move from firm to firm within the "secondary labor market," where skills are non-specific, job tenure is precarious, and few career advantages are obtained by switching firms. In this view, the "primary labor market" includes both career movement within firms (internal labor markets) and movement among high-skilled jobs among firms.

The effects of immigration and of international competition on domestic work organization, including wage rates, is another area of substantial interest. Because of substantial reductions in fertility in the United States in recent decades, immigration and regional migration have become major components of population change. Neoclassical economic theory, which viewed migration as an equilibrating response to differences in prices and wages, has proved inadequate to explain the observed population flows. Recent findings are that expectations about future earnings and changes in nonearnings income, rather than current interregional wage-rate differentials, induce migration, which in turn induces new business investment. Furthermore, research shows that areas with high inmigration also experience high outmigration, and that regional differences in income and unemployment change slowly despite high levels of place-to-place migration. Studies of migration decisions at the household level find that they are based on surprisingly little information. Better longitudinal data on expectations, perceived alternatives, and actual mobility decisions of households is needed to refine and detail the dynamic processes of expectation formation and decision making that influence migration. At other than the household level, data on relevant migrational firm behavior, the role of regional government competition for private investment and development, and the mutual effects between migration flows and the operation of local labor and housing markets are all sparse at this time.

At the international level, recent research has established that migration not only adds a new set of workers into the existing production organization, but it also changes the way in which work itself is organized. Immigrants, especially illegal ones, are much less likely to be employed in large factories than indigenous workers; they work in small, highly mobile shops, at home under a piece-rate system, as sharecroppers in agriculture, or as itinerant wage laborers under the gang system. Old industries like garment and footwear production and new ones like electronics have become increasingly "informalized" through their reliance on immigrant labor. These changes in the organization of work as a result of international migration are of fundamental theoretical as well as policy significance. Research is badly needed to asses the results of the new federal legislation on illegal immigrants in the context of labor markets that rely on immigrant labor.

New Sources of Data on Jobs and Careers

A central limitation facing researchers interested in explaining labor market phenomena is the lack of longitudinal data on individual work histories, work arrangements within firms, and the ways that both change over time that is detailed enough to distinguish competing theories of employment contracting, job search, employment, job design, career patterning, and wage allocation. At present, sizable research investments in dynamic data bases are largely devoted to samples based on households and families.

An example that illustrates the rich returns of these kinds of data is the finding on poverty from the Panel Study of Income Dynamics (PSID), which has been collecting data since 1968 on 5,000 families, chosen to be representative of the national population. (Prior to this research investment, the empirical information base for understanding the well-being of the population came almost entirely from cross-sectional or "freeze-frame" studies, which gathered information from independent samples at one or more points in time.) The PSID results confirmed the cross-sectional finding that, in each year over a 10-year period, about 7 percent of people were in families whose incomes fell below the line. But the PSID data showed that nearly 25 percent of the sample fell below that line in at least 1 of the 10 years, approximately 5 percent during 5 or more years, and approximately 3 percent during 8 or more years.

The PSID results made it possible to distinguish between people who are temporarily poor and those who are persistently poor and to assess the size and character of each of these populations. More generally, the PSID data showed that only about half of the best-off Americans were also best-off 7 years later and only half of the also worst-off were worst-off 7 years later. The major determinants of such changes in income level were transitions in marital status and similar familial events. Such a precise—and unexpected—finding of the relatively large degree of movement that exists within the seemingly stable American income distribution could not have arisen from earlier cross-sectional statistics of income distribution. Now, of course, on the basis of the new knowledge, new cross-sectional studies may be developed to confirm and extend such findings with retrospective queries.

To achieve a similarly enriched picture of job and career dynamics in the context of the employers and firms that produce income and invest in new technologies—in contrast with the households that consume them—requires longitudinal research data comparable in scale. Longitudinal data for large representative samples of firms, jobs, and workers would allow researchers to see how external forces on organizations, including business cycles and attitudinal and technological changes, compare with factors inside organizations, including managerial structures, promotion practices, and compositions of work forces, to shape hiring and recruitment, the design of jobs, and career outcomes. In addition, data could be obtained on decision making that affects employment in the face of sometimes rapidly changing technologies, contractual arrangements with employees, suppliers, and customers of firms, and organizational perceptions, politics, and cultures in firms over time. Such data would permit much sharper empirical tests than have so far been carried out concerning theories of job segregation, wage inequalities, unemployment, productivity, and organizational dynamics and might lead to completely new knowledge about the nature of work and organizations.

A second source of potentially rich data lies in historical knowledge on such matters as how the composition of the labor force responded to past immigration and how the nature of work was transformed by changing technologies and organizational structures in earlier periods. While good longitudinal and comparative data on work arrangements in the past are hard to find, researchers have recently identified several large-scale sources that could significantly enrich historical understanding of work and careers. Several large corporations have maintained detailed data describing employees' job histories over many decades: these records, which are classed as inactive and are no longer of any practical value to the companies, represent a largely untapped source of data for researchers to assess how changing technologies, organizational structures, and labor market conditions affected job design and workers' career outcomes in an earlier era.

Such archives have potential value not only for the study of organizational change and internal labor markets but also for the study of industrial science and technology (discussed in Chapter 4). Some work has already begun in these directions. This work would be well served by a general initiative to develop joint public/private sponsorship of data development and analysis projects to convert an appropriate sample of archives of major U.S. corporations into social and historical research centers or repositories.

During the 1930s the U.S. Employment Service began gathering data on the staffing patterns, promotion ladders, and job requirements of various establishments. Until the program was eliminated several years ago, these data were collected throughout the country (for some organizations, at more than one time) in order to prepare the *Dictionary of Occupational Titles* and other government publications. Microfilm or original documents exist in Washington, D.C., and in the program's central repository in Raleigh, North Carolina. Machine-readable versions of these files would provide researchers with invaluable longitudinal and comparative information on the organization of work and opportunity in American industry over the last half century. Researchers have already converted a very small subset of these data (for example, California enterprises analyzed since 1959) into machine-readable format, developing coding procedures that could be used in a larger effort.

Other data sources in government records might also be useful for studies of work and careers. However, the need to protect the confidentiality of individual respondents places important restrictions on data about organizations that are in the files of the Census Bureau, the Internal Revenue Service, and the Bureau of Labor Statistics, among other agencies. And because of the growth of extensive privately held computer records regarding individuals, households, and firms—such as files maintained by business information companies, direct-mail firms, political action committees, health insurance consortia, credit bureaus, and the like—and the availability of increasingly sophisticated records-linkage software and fast, powerful computer workstations and supercomputers, federal agencies have become even less willing to make edited data accessible to researchers, even after identifying information has been deleted. In fact, there is no instance on record of a qualified scientific researcher using such records-linkage possibilities to identify individuals, much less using such information inimically; however, researchers do share the concern that less benign interests may exploit public-access files in ways that would not be acceptable either to those who have disclosed the information or to the agencies that have ultimate fiduciary responsibility for it. Given this situation, it is worthwhile to explore arrangements that might provide access to disaggregated data files (microdata) on firms and other kinds of organization without unacceptable risks to confidentiality. For example, contractual arrangements for temporary use of screened files by qualified researchers, providing penalties against breaches of confidentiality similar to those that bind government employees, may be an appropriate and acceptable solution.

Questions about individual career patterns that cannot be easily addressed by organization-based or job-based samples—in particular, occupational aspirations, mobility between organizations, and geographic migration—may best be addressed through work-event histories that are based on existing longitudinal panels, such as the Panel Study of Income Dynamics, the National Longitudinal Survey of Labor Market Experience, and the Survey of Income and Program Participation.

There is no simple formula to determine what kinds of research questions can best be answered through new data collection efforts rather than through concerted efforts to gain carefully protected access to file data. In many cases the two resources are complementary. In order to advance this field of study rapidly and efficiently, the agencies involved and representative researchers should work to formulate systematic long-term investment plans for the data needed for the research opportunities discussed above.

OPPORTUNITIES AND NEEDS

Three kinds of work are important to advance the understanding of decisional, allocative, and organizational phenomena: (1) theoretical analyses of choice, information, incentives, and behavior in markets and other organizational contexts; (2) empirical studies, especially of policy-relevant issues, using panel and longitudinal data on organizations and individuals in organizational context; and (3) refinement and extension of laboratory and field experiments to inform both theoretical and policy questions.

In spite of many past advances, knowledge about choice and allocation is still fragmentary. Experimental work has been limited to a few topics, and observational studies in policy settings have been limited by the availability of data. The advances do suggest, however, that continued research along the lines already initiated is fully warranted, and can be expected to generate valuable new knowledge in such diverse areas as regulatory and legislative reform, the design of financial markets, job segregation and wage gaps, and assessment of the implications and effects of corporate mergers and takeovers. We recommend a total of $56 million annually to support research on these topics.

Certain themes are virtually certain to continue in the next decade: the relationship between information, incentives, and the performance of organizations and market systems; the constraints imposed by human and technological limitations on information processing; the effects of divergent goals and dispersed information on the design of organizations; and the formal properties and purposes of contracts over time. New knowledge can be expected from further development and application of theories of individual decision making under conditions of uncertainty, as it arises from incomplete information about the environment and about the motivations and decisions of other actors in the situation. Important work can also be expected on theories of the mechanisms of collective choice, including strategic behavior, the manipulation of agendas, and the possibility of self-enforcing provisions against cheating. Such research will involve all three kinds of research work: theory development; empirical studies, including the collection of longitudinal data, particularly when repeated interactions are the issue; and laboratory and field experiments. For the study of mechanisms and institutions that promote organizational durability, flexibility, and effectiveness (as well as their opposites), a mix of historical, demographic, and ethnographic research methods are needed.

Much of the most valuable research in the topics covered in this chapter has been supported entirely by traditional investigator-initiated grants, and they can be expected to continue to yield rich results. We therefore recommend a substantial expansion of investigator grants, in an annual amount of approximately $20 million. While, overall, new equipment needs are not large by comparison with some other kinds of research, improved computer hardware and development of advanced software for investigators are essential, and we recommend, accordingly, approximately $4 million annually above current expenditure levels for these categories of equipment and support.

We are especially concerned with the diversion of young research talent at the postdoctoral level away from research careers, due to the attractiveness of opportunities in nonacademic careers and the paucity of postdoctoral research positions. We therefore recommend that an additional $5 million annually be added to the support of postdoctoral research fellows and an additional $1 million to predoctoral fellowships.

One of the most powerful devices for encouraging deep and rapid theoretical research has been the fostering, on a continuing basis, of networks of individuals working on closely related issues. The research at the Cowles Commission at the University of Chicago in the 1940s and 1950s offered many examples of the major breakthroughs that can occur if like-minded scientists interact on a regular basis. Much of the recent work on incentives and information has come from initial breakthroughs made by a group of researchers from across the country and abroad, meeting in regular colloquia twice a year. One or two week-long conferences during the academic year and a 2- to 4-week summer workshop, coupled with some resources for graduate students or postdoctoral trainees, have proved able to encourage very rapid breakthroughs on well-chosen subjects. Expenditure for such research workshops should be expanded with an additional $3 million annually.

One of the most important developments over the past two decades is the emergence of controlled experimentation on decision making and the design of market and other types of organizations for allocating resources. Even a simple market transaction is governed by many rules and understandings about property and contract, the value and stability of money, and credit. Political institutions of elections and representation are likewise nestled in a web of rules and procedures meant to regulate the process and produce recognizable and accepted outcomes. These processes have traditionally been studied in real world settings, but successful efforts have been made to bring the study of organizational and market behavior into the laboratory. In a number of cases, the success in doing so without falling into the trap of losing or distorting the phenomenon of major interest in the attempt to isolate it has been striking.

To investigate the importance of rules and procedures, laboratory experimenters have people engage in imaginary but familiar transactions. Experimenters can systematically change the rules of the transaction game, varying the procedures, incentives, information, or objectives given different groups of subjects. The terms of the transactions, for example, may be defined as simple one-shot bartering, as auction or bidding situations, as short-term, high-risk situations, or as long-term, predictable relations between trading partners. The results indicate that these properties are powerful determinants of traders' preferences and show in detail how they can be expected to work. Laboratory experiments, although far from a substitute for field research—which, among other things, is needed to check on the idealizations introduced in the laboratory—constitute an efficient complement that permits much greater control over research variables and overcomes the need to wait for many events to occur in the real world in order to test every plausible hypothesis.

Experimental work is critical to theory development. Experimental methods require specification of the detailed structure of the processes presumed to be operating in the markets or other forms of organization under study. The impact of the new data on theory has also been dramatic. Many basic principles and assumptions have come under close examination, leading to new theories and important revisions of older ones.

Application of experimental methods requires some long-term investments. First, support is needed for the development of additional laboratories, both for equipment, space, and communication devices such as interactive computers, and, more important, for professional staff who can develop software and maintain and improve hardware. We believe that the establishment of new laboratories and the improvement of existing laboratories requires an increased annual expenditure of $2 million.

Second, theorists from a variety of disciplines must be able to participate in the design, evaluation, and interpretation of experiments. The theoretical issues are often so detailed or subtle that sustained communication is necessary to design appropriate experiments. The phenomena of interest often require input from several disciplinary sources so that the emerging set of principles can find use in applications. Colloquia, released time, and provisions for visiting scientists are needed.

Moreover, training in experimental methods must be expanded. A major strength of the experimental method is the opportunity for different researchers to replicate results. Replication necessitates standardization of procedures and methods. Such standardization has been facilitated in other laboratory sciences through decades of teaching laboratory methods in high school and college. Experimental methods to study markets, contracts, organizations, and agent behavior have not had the advantage of such large-scale background support. Training is needed for researchers to learn the procedures of laboratories where experimentation is being conducted, replicate the original results of others, and thereby consolidate scientific advances while gaining high-quality experimental skills. The additional training can be effected in part by postdoctoral fellowships and opportunities to spend periods of a month or two at existing facilities. All of these ancillary activities—interdisciplinary and interinstitutional collaboration in the design of experiments, periodic visits, and training in experimental procedures—could be sustained by a new program of experimental centers at $4 million per year.

This chapter identified specific areas where new kinds of empirical data are needed. Data on expectation-formation at the individual level, for example, possibly including programs of laboratory experiments, would help isolate the causes of departure from rational-expectations hypotheses that are observed in financial markets. Study of rigidities underlying unemployment at the macroeconomic level require detailed data on contract structure and the stickiness of price adjustments. Chronological data on collective bargaining and arbitration, possibly collected in field studies in which the arbitration rules vary, would permit much more accurate assessment of the role of rules and precommitments in successful collective bargains.

This chapter also singled out two types of longitudinal research that promises significant knowledge. First, longitudinal data are needed on the behavior of firms, particularly promotion practices and the trickle-down of vacancies, procedures for evaluating and rewarding performance, and the nature and extent of on-the-job training. These data need to be matched with panel data on workers. Detailed information about contracts with attention to the provisions for wage and hours adjustments and layoff rules need to be collected to assess the impact of economy-wide and firm-specific risk on the welfare of workers. The incidence and effects of multiyear labor contracts with unions should be studied in connection with the observed stickiness of wages. Second, longitudinal data on the organizations as such—in contrast to individuals or contracts—are required to gain a deeper understanding of their dynamics and their strategies of decision making; their response to changes in the economic, political, and legal environments; and their strategies for survival, expansion, and change in general.

Collecting longitudinal data is as expensive as it is important. The initiation of a number of appropriately designed large-scale longitudinal core projects, sustained over the requisite multiyear period, with solid support for archiving, documentation, dissemination, and technical and analytic assistance to users, will require an additional annual expenditure of $12 million. For maximum benefit of such expenditure these data collections should become the core to which are attached research projects that involve other methods, particularly ethnographic studies of the workplace and of occupations, network analyses of job opportunities, and field experiments.

A necessary complementary strategy to collecting new data is to cultivate, supplement, and disseminate existing research data and other potentially valuable data files more thoroughly. To do so requires establishing more effective, better supported lines of communication among academic research centers. An example of such communication is specialty-area computer networks, which have been established largely in psychological and human-developmental areas of research. This strategy also requires the establishment and maintenance of better lines of communication between researchers and the relevant data-collecting administrative agencies and private firms, to ensure that records assembled for purposes other than science or research can be made as useful as possible to the scientific community. We estimate that the total range of appropriate efforts to improve access to the most useful data that now exist in a series of academic research centers, government agencies, and private sources will cost approximately $5 million annually, of which $1 million should be especially directed to the exploration and cultivation of private record centers (such as insurance clearinghouses) and unused corporate and local government archives for research purposes.

4

Institutions and Cultures

4

Institutions and Cultures

In many respects the preceding chapters may be regarded as moving from the more microscopic aspects of human existence to the more macroscopic. Chapter 1 emphasized research on individual thought and behavior, paying particular attention to biological and psychological determinants and mechanisms. Chapter 2 focused on motivational determinants and mechanisms in individual behavior, as well as research on the constraining and determining effects of social interaction and social arrangements with respect to behavior. Chapter 3 took up organizational arrangements, especially markets and political and occupational systems, and examined how information, incentives, and other features shape these arrangements.

In this chapter we move further toward the macroscopic and consider research on institutions and cultures, which are those features of social life that serve as the bases of the organization and integration of entire societies. Although the analysis of institutions and cultures is associated primarily with the fields of sociology and anthropology, aspects of these phenomena also relate to and are found in the research of political scientists, historians, geographers, legal scholars, and economists as well.

Research on institutions and cultures has long been an active field, and there is now a vast array of ongoing work. This chapter begins with the most basic questions of human beginnings: What have been the salient characteristics of human evolution that have led to that special type of social bonding that is described as human society? New archaeological studies as well as the comparative study of primate and human societies are now adding major insights on this central question in evolutionary theory. Another major field of study in this area is demography—the study of population dynamics as affected by fertility, mortality, and migration—and this chapter focuses on the institutional and cultural dimensions of both fertility, especially in developing societies, and migration, especially in the United States.

One question has been the preoccupation of nearly two centuries of theoretical thinking in the social sciences: How can the profound institutional transformations that have been associated with the great commercial, industrial, and democratic revolutions in modern Western history be understood and explained? In recent decades this question has been extended to include the study of developing nations as they struggle with change. Study of these institutional transformations has gone under a number of labels, none of them completely satisfactory, but for purposes of convenience we will call it the study of modernization. Within this large and active field, the chapter focuses on the family and religion. It considers a range of research that has significantly altered understanding of how these institutions have changed and continue to change and what place and role they have in modern society.

Science and technology have been among the key features of those institutional transformations and a subject of intense recent research: What cultural, institutional, and organizational features of society are essential for scientific knowledge to rise and grow? What are the conditions that determine whether scientific knowledge will be applied, that is, implemented as technology? And, then, what effect does technology have on communities and institutions and on the quality of life in general? These questions are studied both by historians of science and by other social scientists. The chapter also takes a look at the behavioral and social sciences themselves—also the children of modernization—and on the relations between social science knowledge and public policy.

Lastly, the chapter deals with the most macroscopic level of all, the world as a whole, where the stress is on systems of relations among societies. Some of the developing lines of research are on internationalization—the increased economic, political, and cultural involvement of nations in one another's affairs—and on the special and timely topics of international security and, more generally, cooperation and conflict among nations.

THE EVOLUTION OF HUMAN SOCIETY

Every society has developed beliefs about the origins of the world and the nature of things within it, including beliefs about human nature and society itself. Since written records are a relatively recent historical phenomenon, the way in which society actually emerged lies in the realm of prehistory (prior to written records). But a wide range of scientific techniques are now making possible an increasingly complete narrative of origins based on the interpretation of material evidence. This evidence enables researchers to look back with increasing clarity to the world of thousands—and millions—of years ago and to visualize the evolution of humans and human society. The outlines of that evolution can be divided into six stages:

Time Period	Developmental Stage
Between 8 million and 4 million years ago	Divergence of the hominid line from African apes
By 4 million years ago	Evolution of bipedal gait Habitation of African savannahs
By 2 million years ago	Larger than ape-sized brains Simple stone tools Butchery marks on animal remains More than one species in the same region (*Australopithecus* and early *Homo*)
By 0.5 million years ago	More complex stone tools Spread into tropical and temperate Eurasia *Homo erectus*, then archaic *Homo sapiens*

Between 50,000 and 30,000 years ago	Loss of massive muscularity
	Emergence of art, technical ingenuity—an
	  "information explosion," including
	  extensive stylistic differentiation and change
	Probable increase in population densities
	Spread to high Arctic, Australia, and,
	  later, the Americas

Between 10,000 and 5,000 years ago	Agricultural transition Settled villages and towns Food storage, cultivation, herding Great increases in population densities Accumulation of wealth, concentration of power

The first two stages saw the development of biological mechanisms similar to those governing the evolutionary branching of many primate and mammalian species. The last stage, involving the beginnings of farming, witnessed the appearance of populations with all the distinctive mental, social, and technological potentials of contemporary humans. The intervening stages, during which changes in biology, behavior, and culture occurred, are the subject of frontier research on the dynamics of social change.

Social Organization in Prehistoric Times

Direct evidence on the social organization of early human groups is limited, but some tentative conclusions can be drawn. Many archaeological sites dating from 2 million years ago to the beginnings of farming 10,000 years ago are of similar size, and they suggest the existence of day-to-day social groups usually of 10 to 30 individuals. Population densities over large areas were very low.

The firmest early evidence of household or family units within sites dates from 35,000 to 50,000 years ago. Signs of wider and more diversified networks also appeared at this time, as did evidence of long-distance exchange of items such as seashells and obsidian and other prized stones. This stage is also marked by evidence of gatherings at ritually important places, the development of regional stylistic traditions that reflect some kind of ethnic identity, and increases in the volume of organized information being generated and transmitted.

In the 1960s, field investigations of nonhuman primates took some observed baboon societies as the model of early human society. In line with that model, the societies of human ancestors were depicted as hierarchial, with males being competitive and aggressive and females being passive and nurturers of the young. Subsequent primate field research has dispelled a number of these inferences: it has shown that primate behavior varies enormously, both within and among species, and that simple generalizations about sex differences in nurturance, social competitiveness, and passivity in females cannot be sustained. Contemporary field studies, for example, reveal that in various species, females compete as intensely as males and often actively choose their mates. At the same time, in some species, including baboons, males develop long-term bonds with females and participate in infant care.

Along with new evidence and changing views on primate behavior, there are indications that major changes in human reproductive physiology may have occurred relatively recently. There is some evidence that Neanderthal females may have carried their fetuses to 11- or 12-month terms. Other evidence suggests that, about 10,000 to 15,000 years ago, when some humans became less nomadic, birth spacing was reduced, which, in turn, was a factor in causing the population increase associated with the beginnings of farming.

With increased wealth in the form of investments in capital improvements to land, signs of warfare appear. The oldest archaeological evidence of organized armed warfare, as contrasted with incidental skirmishes, is a 12,000-year-old cemetery in the Nile Valley. Although signs of warfare follow rather than precede the development of farming, fighting as such goes further back. Present-day monkeys and apes, especially males, fight, as do modern humans; and there is no reason to suppose that human ancestors were different. The question is rather one of scale, intensity, organization, and the use of weapons. Fieldwork on chimpanzees indicates that they may share with humans the dubious distinction that organized coalitions of males engage in lethal intergroup aggression to achieve territorial gains. Studies are needed to confirm the finding and to determine the evolutionary and social context of these patterns.

Food, Tools, and Home Bases

At some stage in evolution, humans joined the class of animals that do not simply consume food on the spot but carry it to central places, called home bases, where it is shared with the young and other adults. The use of home bases is a fundamental component of human social behavior; the common meal served at a common hearth is a powerful symbol, a mark of social unity. Home-base behavior does not occur among nonhuman primates and is rare among mammals. It is unclear when humans began to use home bases, what kind of communications and social relations were involved, and what the ecological and food choice contexts of the shift were. Work on early tools, surveys of paleoanthropological sites, development and testing of broad ecological theories, and advances in comparative primatology are contributing to knowledge about this central chapter in human prehistory.

One innovative approach is to investigate damage and wear on stone tools. Researchers make tools that replicate excavated specimens as closely as possible and try to use the tools as the originals might have been used, such as wood-cutting, hunting, or cultivation. Depending on how the tool is used, characteristic chippage patterns and microscopically distinguishable polishes develop near the edges. The first application of this new method of analysis to stone tools that are 1.5 million- to 2 million-years-old indicates that, from the start, an important function of early stone tools was to extract high quality food—meat and marrow—from large animal carcasses. Some of the earliest tools were also used for shaping wood and for making digging sticks and spears. Fossil bones with cutmarks caused by stone tools have been discovered lying in the same 2-million-year-old layers that yielded the oldest such tools and the oldest hominid specimens (including humans) with larger than ape-sized brains. This discovery increases scientists' certainty about when human ancestors began to eat more meat than do present-day nonhuman primates. But several questions are still unanswered questions: How frequently did meat-eating occur? To what degree was meat acquired more by scavenging than hunting? What were the social implications of meat-eating patterns?

New analyses of animal remains from the stone age are now under way. An important question to be addressed by field studies is how the feeding, ranging, and social interaction patterns of human beings who acquired food by hunting and gathering compare with those of primates and nonhuman carnivores. Some studies need to span the life cycle of identified individual animals, particularly of apes and other species for which there is evidence of repeated strategies and accumulated experience that amount to a protoculture.

Until recently few such studies were undertaken, and they were qualitative. Now optimal foraging theory and other rigorous conceptual frameworks are being used in gathering data on the properties of nonagricultural human and primate foods, notably their spatial and seasonal distribution, associated acquisition and processing costs, energy and nutrient returns, and problems caused by toxins and secondary compounds. A few exploratory field projects have begun on wild tubers, meat and fat in scavenged carcasses, and such problems as coping with tannins and their effects on food choice. Of related interest are claims that control over fire may go back 1.5 million years, much earlier than commonly thought; fire control had definite implications for food selection and gathering behavior. The very existence of extended human social systems may have been determined by the ways prehistoric people learned to exploit widely dispersed, high-quality, portable food products—meat, marrow, large tubers, and, much later, grain.

Evolution of Language

Language, the most important component of culture, is also the most difficult subject to study in an evolutionary framework because there are no living protolanguages and speech does not fossilize. The initial stages of brain expansion and stone-tool manufacture, both suggestive of language facility, began about 2 million years ago. But there is also some evidence of a relatively recent change in the structure of the human vocal tract. This change coincides with the loss of muscularity that distinguishes anatomically modern humans, with modern-size brains, from the Neanderthals whom they replaced about 30,000 years ago. Further assessment of this biological history, including a more thorough investigation both of vocal tract anatomy and brain structure, is needed.

The social and ecological factors that gave rise to the evolution of language are unclear. Competing hypotheses about what made language-like communication useful range from considerations of foraging strategy to problems of mating and infant care. Information sharing could have played a critical role in organizing home-based foraging, in cooperative hunting, in resolving intergroup conflict, and even in maintaining stable mating and provisioning relationships between pair-bonded mates, as well as between different mating groups. These and other hypotheses can only be tested by pursuing archaeological evidence on dwelling and foraging patterns (for example, home bases and tool caches) and integrating these findings with more thorough observational studies of the dwelling and foraging strategies of modern foraging peoples and other primate species. The study of language origins places strong demands on the capacity of scientists to integrate paleontological, neurological, ecological, and behavioral studies into a coherent picture and, to put together the results of biological and historical field methods, laboratory analyses of material gathered by radioisotopic or microscopic instruments, data from experimental techniques, and well-articulated theories of behavior in the wild.

But modern methods can only illuminate the past if there are systematic and well-preserved artifacts available for study. There is a compelling need to protect museum-based research collections against physical deterioration and dispersal and to cultivate international arrangements for scholarly access to worldwide observational and archaeological field sites.

DEMOGRAPHIC BEHAVIOR

In one sense demography is the study of phenomena that are primarily biological, such as fertility and mortality, which in the aggregate can be analyzed with sophisticated mathematical models and statistical techniques. Yet these phenomena cannot be understood simply as biological processes because they are strongly influenced by social and cultural factors. The major task in studying demographic behavior is to unravel and analyze the complex of causes behind rates of fertility, marriage, migration, morbidity, and mortality. This task calls for a variety of sources of information, including official demographic records and surveys, fine-grained observations in field situations, and careful sifting and analysis of historical records. In this empirical work it has been important to compare detailed data from surveys with theoretical and mathematical work and to compare cross-sectional information from such surveys with personal and institutional longitudinal data. Future understanding will depend largely on multifaceted and sustained research efforts, especially in regions of Africa and Asia where cultures and institutions very different from those in the United States present the greatest challenges to understanding the dynamics of demography.

Fertility and Lactation

Unlike animals, humans control reproduction (directly or indirectly) through cultural mechanisms, such as celibacy outside marriage and abstinence within it. Prolonged breastfeeding is another instance of this kind of control. The physiological effects of breastfeeding are now well understood: the mechanical stimulation of an infant's suckling triggers hormonal mechanisms that delay the mother's return to normal fertility. Complex simulations and statistical analyses of these physiological processes have increased the precision of knowledge as to how this effect varies according to the frequency, intensity, and duration of breastfeeding—which are determined by cultural forces. Those forces, manifested mainly in family and community norms, determine how and how long an infant is nursed as well as when and in what manner supplementary feeding begins. Broad social and economic conditions, including work and the mother's social expectations and health, affect whether and how much she breastfeeds. This web of biological, personal, and cultural forces determines whether the mother or someone else nurses and when she shifts the child to other foods, which in turn affect the survival and health of the child. If a nursing child dies, as frequently happens under conditions of poverty and disease, normal fertility returns, thereby increasing the probability of another pregnancy. And multiple pregnancies may have detrimental effects on the mother's health.

In some societies breastfeeding is explicitly regarded as a device to lengthen the spacing between births. Fewer births increase the food supply for the existing children, thereby improving their health. In addition, breastfeeding may be accompanied by abstinence from sexual intercourse. Breastfeeding plus abstinence can result in birth spacing of from 3 to 5 years. As societies change, abstinence or breastfeeding or both may be abandoned for various reasons, with a consequent rise in fertility, an increase in infant mortality (due to the lack of enough nutritious and uncontaminated breast milk), and a decline in the health of women through repeated childbearing.

Another cultural force that can affect birth spacing is the value that cultures place on children. Male and female children are often valued differently, and such different valuation may affect the care that an infant receives. Less than a century ago, the survival of female children was lower than that of males in regions of Western countries where females were believed to contribute less to the family economy than males, and this effect can also be observed currently in non-Western countries.

Population Change in Developing Countries

Fertility in developing countries has become and will remain a major focus in demographic research because of policy concerns with rapid population growth. However, scientific interest in the relationship between population change (especially growth) and human life in general (especially economic well-being) dates back to the eighteenth century. Research has shown that in modern society large family size is generally detrimental to the health and well-being of families and their members; this effect exists across different economic status. Families that limit their size benefit as families, as do their members as individuals, on a variety of measures. However, the effects of families that limit their size on families that do not, the effects of family size limitation on society as a whole, and the broad social and political consequences of population growth and decline are little understood. These issues will occupy an important place in demographic research over the coming decade.

Many advances have been made in sorting out the biometric features of the fertility process, and further gains in the precision of both measurements and theoretical formulations can be expected. A key to research advance is the careful analysis of specific cases of fertility declines in their social context, which includes family economics, local and national administrative systems, cultural change, and deliberate governmental efforts aimed at reducing fertility. For example, a series of empirical studies have demonstrated how ideological systems imply particular emotional satisfactions in childrearing: in some systems, there are accepted or preferred substitute goods or services; in others, there are not. In the latter case, fertility does not drop off sharply in response to reduced mortality and improved old-age economic security as it does in the former case.

There are several competing theories about the efficacy of various factors in reducing fertility: that declines in fertility result from the elimination of "unwanted" childbearing; that structural factors affect couples' cost-benefit thinking about having children; or that the decline of fertility is a response to the diffusion of Western ideas through the developing world. The recent work of the World Fertility Survey and the European Fertility Project lends some support to the diffusion theory, but the search to uncover the combination of factors underlying fertility decline continues.

An impediment to research on fertility in developing countries is the inadequate registration of births, marriages, and deaths. Lacking these kinds of data, researchers have devised indirect strategies for estimating levels and trends of fertility and mortality. Questions that can yield such information are included in censuses or surveys. Estimates of levels of fertility and other unobservable demographic variables are also extrapolated from known population features, especially age distribution and growth rate. These methods are valuable for tracing demographic trends not only in the developing countries but also in the United States, where phenomena difficult to observe in aggregate data, such as the durations of marriages, have been built into estimation models. The growth of estimation methods has been so rapid that a 1983 compilation of state-of-the-art methods by the United Nations is already obsolete, and continued development and testing of new estimation procedures is essential for increased knowledge.

Research on urban growth in the developing world is also hobbled by unsatisfactory data. Because of the rate of growth and the fact that most censuses are conducted only every 10 years, data on city size is out of date by an average of 6 years and data on urban rates of growth by as much as 11 years. Urban planning policies that are based on such outdated data are vulnerable and can result in costly mistakes. Yet most developing countries do not have the resources to conduct more frequent censuses, to institute population registration systems, or to develop regional population estimation capabilities, like those developed successfully in such countries as the United States and Great Britain. Indeed, some countries in Africa, such as Ethiopia, Nigeria, and Zaire, are finding it difficult—because of political conflict and economic strains—even to conduct regular censuses. One ingenious, new way to compensate for the lack of data is to use remote-sensing (satellite photo) techniques, which can provide estimates of the total population of urban areas in developing countries that are practically as accurate as their on-the-ground censuses. Development of this method will require substantial technical support and access to the kind of data gathered by military and intelligence satellites.

Much of the knowledge about the biosocial aspects of demography in developing countries derives from the International Centre for Diarrheal Disease Research in Bangladesh (formerly the Cholera Research Laboratory). This facility maintains an up-to-date register of about 200,000 people in Bangladesh, whose demographic characteristics and vital events are recorded and monitored. The Centre has contributed to scientific knowledge of the role of childhood diarrhea in growth retardation and mortality and the importance of lactational amenorrhea as an inhibitor of fertility. It has been the site of the most carefully prepared evaluations of family planning programs and has produced the most accurate demographic tables for South Asian countries, which are used as a model for other populations. The Centre maintains a strong social science emphasis, hosting both resident and visiting demographers, anthropologists, and sociologists. A second institute, the Institute for Nutrition in Central America and Panama, located in Guatemala, has also proved successful, although its staff is less experienced and its functioning has been impaired by Guatemala's political situation. These institutions serve as a model for developing another facility to gain new knowledge from and about the unique environments in Africa, western Asia, and South America. Demographic investigators now rely heavily on secondary statistical analysis of large data files from these areas, and they need appropriate new facilities to gain in-depth, first-hand knowledge that permits the incorporation of cultural, social, and psychological dimensions into their work.

Fertility and Migration in Developed Countries

Fertility research in developed countries focuses mainly on low levels of fertility, now below replacement levels in parts of Western Europe and North America. Overcoming infertility among childless couples is one area of concern. Other research focuses on government efforts to enhance fertility through incentive programs, which, however, have not been very successful. The social and economic consequences of low fertility and the resulting relatively small population cohorts are also major research topics. Countries experiencing low fertility must rely on immigration to expand or maintain their labor force and so must face the social and economic challenges of accommodating immigration. In addition, the new small cohorts face the costly prospect of supporting earlier large cohorts, who will be retired when the former are in their prime working years.

Low fertility in the context of perceived labor shortages encourages immigration. New migrants are generally young and tend to bring with them the fertility patterns of their previous home cultures. For example, in the United

States in the 1980s, 21 percent of all births were among Hispanic and Asian groups, who constituted less than 9 percent of the total population. The characteristics of the Hispanic and Asian immigrants and their preferred destinations often have a significant impact on relative regional growth and demand for public services. Their presence has reraised the great nineteenth- and early twentieth-century issues of cultural and linguistic assimilation, ghettoization, and ethnic discrimination. Explanations and models from those earlier periods, however, may not be applicable: it is not yet known to what extent the processes observed for earlier migrations are general and to what extent they were particular to those immigrant groups. Research must answer questions about the dynamics and prospects of migrant assimilation, the social and economic impact of immigration on receiving areas, the character of the new migratory groups, and the results to be expected from the kinds of reception they receive from communities and municipal governments.

The policy implications of high rates of international migration are evident. If fertility levels remain low, the role of immigration as a means of ensuring population stability or growth is likely to continue. Most recent migration to the United States originates in Latin America, the Caribbean, and Asia. The experiences of these immigrants with respect to schooling, language, and labor-force participation are important in discussions of immigration policy. For these reasons, comparative work is essential on the impact on wage rates of migration from poorer regions, and the effects on capital formation, social welfare expenditures, population growth, and political instability in other developed countries that have had experience with migrants, for example, Sweden, France, Italy, and West Germany.

There are virtually no data on emigration from the United States, and data on net migration to the United States are poor. Consequently, this component of population growth—which may account for half of the total increase—is subject to serious measurement error. In addition, there are scant longitudinal data on the movements of individuals and households within the United States; such data could reveal much about how internal migration decisions are made and how these movements affect regional and community social systems. The greatest improvements in detailed knowledge about both external and internal migration patterns can come from present data bases such as the Panel Study on Income Dynamics, the Survey of Income and Program Participation, and records of the Internal Revenue Service and the Social Security Administration. But the kinds of information on the locations, movements, origins, and destinations of people that these data bases contain are usually not made available for analysis even in modified form because of confidentiality considerations. Inventive ways can and should be devised to make them available for research purposes while at the same time protecting individuals' privacy. (This issue is discussed more generally in Chapter 6.)

MODERNIZATION: FAMILY AND RELIGION

Until recent decades, most investigators of institutional changes proceeded on the assumption that all societies experience similar kinds of change as they modernize. Among the presumed uniformities in the modernization process were:

- Application of nonhuman and nonanimal sources of power to production, development of large-scale enterprises, an increased division of labor, and a resulting increase in economic productivity.

- Movement of the population from farms and villages to urban centers.

- Development of mass-schooling systems, resulting in greatly increased literacy.

- Decline of extended family systems and the rise of the isolated nuclear family.

- Decline of patriarchy and increase in sexual equality.

- Weakening of traditional religion and the corresponding secularization of society.

- Opening of society's stratification system, with a person's place based more on individual achievement and less on inherited familial status.

These assumptions were widely accepted until about 20 years ago, and they guided most empirical investigations of modernization and institutional change; they are now changing substantially. Innovative theories and an abundance of new empirical evidence have deepened the understanding of modernization in general and changes in religious and family institutions in particular.

The Nuclear Family and Social Change

Modernization theory once viewed the family before the commercial and industrial revolutions as patriarchal, based on arranged marriage, extending beyond the procreative couple to include grandparents as well as lateral kin, and gaining legitimacy from the traditions of religion, community, and property. Then, under the demands of geographical and social mobility resulting from commercialization and industrialization, this kind of family was thought to have been replaced by one that was mainly isolated from extended kin, based on romantic love, possessed of greater equality between the sexes and between parents and children, and authorized by a civil, secular contract. Some scholars evaluated these trends as evidence of a serious deterioration of the Western family; others heralded them as a sign of the family's adaptability and its institutional "fit" with other modern institutions. Whatever the evaluation, however, there was general agreement on the main lines of change.

For the past two decades, however, research on a broad multidisciplinary front has criticized, complicated, and enriched this view of the history of the family. First, new evidence has supported the view that the residentially autonomous nuclear-family type came before rather than after other institutional signs of modernization. The nuclear tendency of western European families as they emerged from the medieval period—a tendency discovered as historians and other social scientists applied new analyses and theories to previously unused or recently reconstructed historical archives—provided an environment in which many children were raised in an emotionally supportive atmosphere to be relatively independent, mobile, risk-taking, and nontraditional. These kinds of families played an active part in originating just those economic revolutions that were previously thought to have spawned the nuclear family. Furthermore, the bonds of the extended family did not everywhere perish as a result of modernization. Studies of how people seek advice, support one another, and give and receive credit show that generational and lateral kinship ties remain vital. Modern technology—automobiles, airplanes, and telephones—often help these ties to persist even when families are dispersed geographically.

Recent historical and cross-national research, using methods and ideas developed in tandem with the behavioral and social sciences to reconstruct the decisions of everyday life and their effects, also shows that the family, far from being a simple "victim" of industrialization, sometimes facilitated industrial development. In sites as diverse as early nineteenth-century England, early twentieth-century New Hampshire, early twentieth-century Japan, and mid-twentieth-century eastern Europe, the family has been an important agency in the accumulation of capital, the recruitment of workers, and the coordination of activities in the workplace.

Comparative historical research has also shed important new light on the ways that societies define and structure the human life span. Research indicates that the stages of life associated in large part with age and family status—childhood, adolescence, youth, mid-life, and old age—are not simply biologically given. They appear and are consolidated as a product of economic and educational forces. For example, the modern idea of adolescence as a distinct phase of life crystallized only in the nineteenth century, as the years between 12 and 18 became especially ambiguous: the family had given up much of its direct control over courtship, marriage, and economic training of the young; personal apprenticeship as an initiation into adult work roles became less common; the factory came to be regarded as an unsatisfactory if not evil place for young people; and age-graded secondary schools had not yet become a vehicle for organizing those years. As a result of these social developments, what became known as the adolescent years demanded new social attention and eventually became a focal point for such concerns as urban crime, immorality, and unemployment.

Finally, diverse lines of research have established that increasing equality between the sexes is not, as it once appeared, an inevitable accompaniment of modernization. Rather, it is a phenomenon dependent on complex economic, legal, social, cultural, and political forces. Certainly the ideology of patriarchy has been weakened by subtle and not-so-subtle changes in family roles, by increased availability of affordable birth control methods, by women's increased participation in the wage economy, by the extension of legal rights, and by the political accomplishments of women rights movements. Yet even this myriad of forces, all working in the same direction, has not obliterated sexual inequality, which continues to show a certain recalcitrance to change not envisioned by modernization theorists (see "Jobs, Wages, and Careers" in Chapter 2).

Partly as a result of this enriched understanding of the historical evolution of the family, contemporary research has addressed the interaction between the family and other kinds of economic, legal, and social phenomena. One example is the effects of separation and divorce on women and children. For a long time sociologists have noticed the general relationship between modernization and increasing rates of divorce, and they focused research on the economic, legal, and psychological factors that might cause this increase. More recently, the impact of statutory changes on the economic position of divorced women has become a topic of intensive interest. In many ways the "equality" of the sexes under no-fault divorce has proven to be a mirage; divorce coupled with the failure of ex-husbands to accept their financial responsibilities leaves many women in positions of economic hardship. Research has also shed new light on the range and length of effects of divorce on children. Longitudinal studies, based on following children through years of postdivorce experience, reveal a pattern of fright, bewilderment, and blame lasting up to 2 years after divorce. Thereafter, these emotions fade, except for a minority of children who experience unhappiness, dissatisfaction, and loneliness for 5 or more years after divorce.

Another major focus of research is on the consequences of three contemporary trends: the proportion of women in the labor force, which has reached new heights in the United States and other industrial countries; the persistence of wage levels for women that are consistently and significantly lower than those for men; and divorce rates in the United States and other industrialized countries, which are at unprecedented high levels. One consequence is that the majority of people in the United States whose incomes are below the poverty line are now living in households consisting of or headed by females who are single parents, often receiving welfare benefits. But panel study data show that the typical period of impoverishment and welfare dependency is a temporary one. Welfare recipiency is generally used in the early stages of recovery from the economic crisis caused by the death or departure of a husband, a process that often culminates in finding more remunerative employment, getting married again, or both. Many but not most of the children of welfare families do not themselves receive welfare benefits after leaving home and forming their own households.

Religion, Social Change, and Politics

The scientific study of religion through historical and cross-national research has followed a path somewhat similar to that of the study of the family. One view prevailing until recently, derived in part from modernization theory, was of an overall trend toward secularization: as societies became more modernized, religion tended to lose its force. Reciprocally, religion was associated with more traditional, conservative societies, and religion itself was considered to be a conservative force, maintaining tradition and stabilizing long-term social and cultural arrangements.

A countervailing perception was inspired by the classical early twentieth century analyses of the role of the Protestant Reformation in the emergence of capitalism in the West. It was argued that a key to the dynamism of Protestant societies lay in part in a religious ethic of revealing and doing God's will through worldly work in a calling or profession. Believers were strongly motivated toward instrumental advancement in ordinary spheres of activity, and not, for example, toward monastic retreat from worldly concerns. By the same token, it was argued that other major religions, particularly the great Indian ones, which encouraged withdrawal from the everyday world and its material concerns, could not engender the same historical dynamism.

Studies undertaken in recent years in southeastern and eastern Asia, using methods and data far more extensive and systematic than anything available a half-century ago, have demonstrated that Buddhists and people of other faiths have worked just as hard and as instrumentally as Protestants, and, furthermore, that these Asian societies have contained significant entrepreneurial sectors. The fact that they did not "develop" in the same way as did the West cannot be attributed directly to particular economic values that their religions might foster.

The question of the relationship between religion and social change has also been broadened to include the relationship between religion and politics. This shift was a response not only to intellectual developments within the scholarly study of religion, but also to the recent dramatic surge of politically oriented religious movements in the Middle East, Latin America, and the United States. Some of these movements have been radical in ideology and some have been conservative, but in either case they have been highly activist. Although popular movements were previously studied to some extent, they were not at the center of research about religion and social change. Rather, religion was seen largely in terms of "the church" or other established forms, with popular movements as unusual and sporadic occurrences, or in the modern era, as responses to the pressures and stresses of modernization—outbursts of people who had no other effective alternatives for social and political action. A more sophisticated formulation held that such movements were prepolitical, that they represented essentially political behavior among traditional peoples with no experience in secular political action, and that, under appropriate conditions, they would evolve over time into conventional political activities. Both views agreed that the religious impulse would lose force where more "modern" alternatives become available; however, this hypothesis has not been upheld in most cases.

Popular religious movements are now understood as central to religion and important for politics. Much of the current study of religious movements in Latin America, for example, takes into account the interplay between the established Catholic church and liberation theology movements, considering how the church limits these movements, and how they bring about profound revisions in church policy, practice, and theology. Other contemporary studies of movements in the Philippines, Africa, the Middle East, and the United States emphasize the ongoing and enduring mutual impact of religion and politics on one another.

SCIENCE, TECHNOLOGY, AND PUBLIC POLICY

The relationship between the growth of science, technological development, and social change has always been a core concern in studies of the long-term transformation of Western societies. These relationships are indispensable to understanding how the modern world came to be, the nature of scientific and technological work, and the policy decisions that affect that work. For a long time the main stress was on the technology of mass industrial production, on tracing the revolutionary implications of advances in science and technology for the organization of the workplace, the labor force, and, ultimately, family and community life. While approach has remained strong, recent research has added other areas, particularly the societal role of scientific ideas and of technological products, such as computers and new reproductive technologies, used by consumers directly.

In the past two decades especially, some of the focus on the relations between science, technology, and society has turned in more policy-oriented directions. Controversy has arisen over the quality and the dissemination of knowledge about environmental risks and economic and human costs associated with agricultural, biomedical, and industrial advances and over the nature of public control and regulation intended to provide the benefits of scientific and technological progress while minimizing the hazards. These controversies about the social responsibility and public accountability of scientists and the uses of new technologies are providing the basis for new perspectives on the research community and on managing technological risks. These new perspectives depend in part on advances in behavioral and social sciences research.

The Shaping of Technology

In earlier periods, historians and other social scientists tended to regard technology primarily as shaping, rather than as shaped by, social organizations and institutions. Research has led to important findings about the reverse side of the technology-society coin: the institutional shaping of technology. On three different levels, the study of social institutions constitutes a crucial link in understanding this relationship. First, the invention of technology takes place mainly in institutions for science and research. Second, the application of technology is influenced by institutional factors, such as the military aspect of the public sector, the organization of commercial firms, and the family and church in the private sphere. Third, the regulation of technology has become an important issue for political institutions.

An accumulation of historical studies have now analyzed the way that dominant forces in society, including cultural settings, values, ideologies, and political and economic structures conditioned the development and introduction of new technology and the emergence of entire industrial systems. A useful early (1960s) case study analyzed the evolution of the Wisconsin dairy industry, including the role of chemical and bacteriological advances and the activities of the University of Wisconsin, the State Experiment Station, and agricultural extension services. A more recent study of the invention and evolution of integrated electric light and power systems in the United States, Great Britain, and Germany pioneered a fully faceted, cross-national approach. The research challenge at present is to develop comprehensive comparative studies of other indispensable technologies. With regard to the twentieth-century history of the exploration, exploitation, and regulation of the radio frequency spectrum, for example: How and why has the spectrum been allocated among civilian and military users and between different commercial users? What has been the relationship between the regulatory environment and technological development?

In the United States, the federal government, often in conjunction with state organizations, has played a major role in fostering innovation (as the agricultural case shows). The military has been responsible, especially since World War II, for stimulating a great deal of research and development. Thus, it is essential to explore the broad range of activities through which the armed forces have promoted, coordinated, and directed technological change and affected the course of modern industry, including the vexing question of evaluating spinoffs from military to civilian applications. On the nonmilitary side, much technological research and development has been produced by a relatively small number of corporations that have institutionalized multistage processes from basic research through applied research, pilot plants, and intensive coordination with corporate marketing departments. Different studies have stressed, respectively, the technical and managerial aspects of new technology, the influence of different cultural settings on the development of technology, and the role of workplace processes in technological change. This multistage process of corporate research and development is among the most important topics for research on the history of science and technology, business history, and economic history. It is intimately connected with the issue of the ability of the United States to compete successfully in world markets.

Comprehensive historical studies of technological development have now become easier because some of the major companies that pioneered in research and development have created archives and hired historical researchers or provided funds for outside researchers to use them. Knowledge in this field will also be advanced with the development of data sets that cover technical manpower, funding, training, employment, and biographies of technological innovators, as generalized resources for research in the field. For all such data, as with government data, it is essential to strike the right combination of free research access and appropriate protection for privacy and confidentiality.

The Development of Science

In the social science study of science, two questions stand out: What social arrangements accelerate the pace and influence the directions of scientific development? Why and how do these arrangements work? Although some people argue that the vitality of any branch of science depends in a virtually linear fashion on the resources invested, careful quantitative cross-national study does not support that claim without modification: money has been a necessary but not a sufficient condition for scientific progress. One cross-national study of the personnel, funding, institutes, and productivity of physics in universities in the United States and western Europe at the turn of the century discovered that financial investment and manpower, calculated as fractions of total national spending, were very nearly the same in France, Germany, Great Britain, and the United States; however, productivity, as measured by numbers of papers published in the leading journals per dollars or per person, differed greatly, from a high for Germany to a low for the United States. American physicists at the time were claiming that poor funding accounted for their relatively low international standing, but clearly that was only part of the story. Subsequent research suggests that while a supportive base of resource support is essential, the intellectual and social environment sustaining scientific discovery also depends heavily on who designs and controls teaching institutions, specialized research centers, and professional societies and journals. The quality of science is ultimately a product of human and economic resources mobilized within social traditions and cultural contexts that enhance and legitimate scientific research and debate. Recent investigations involving close qualitative scrutiny of the conduct of scientific work in the laboratory, lecture room, seminar, and one-to-one discourse are enriching the understanding of how scientific facts and ideas are brought to light and accepted or rejected.

To further test and consolidate these kinds of findings, comparative quantitative research is essential. Quantitative studies can provide data concerning growth rates in the respective sciences, the differential productivity of practitioners and institutions, and differences in recruitment into science. Such quantitative results research must then be informed by the findings of qualitative cultural and institutional studies. For example, numerous studies have shown that certain minority groups in the United States—Jewish Americans historically and later Asian Americans—participate in science to an extent far greater than their proportion in the population, and the opposite is true for other minority groups and for women. Explanations of the uneven participation of women and minorities in science requires study of the cultures of the groups themselves, their histories and opportunities, and the policies, practices, and cultures of educational and scientific institutions.

How is the scientific community organized? How are research priorities established? How do these affect what work gets done? Recently it has come to be acknowledged that a great deal of contemporary scientific and technological research is the creation of research teams, either at specific locations or as dispersed "invisible colleges." A lone scholar can readily study Galileo; the lone researcher whose subject is the Brookhaven National Laboratory or genetic engineering faces a more formidable task. What teams of scientists, engineers, technicians, and managers have created will take teams of historians, sociologists, economists, and other researchers to analyze. The study of modern science and technology requires long-term collaborative interdisciplinary effort, and current funding patterns and scholarly practices, based on individual projects over limited time periods, have not been conducive to such long-term collaborative effects.

Behavioral and Social Sciences Knowledge and Public Policy

An important trend in research on the development of science is the inclusion of the behavioral and social sciences as subject matter; that is, study of the emergence and institutionalization—including modes of recruitment, career incentives and patterns, and influences on intellectual preoccupations—of the modern disciplines and professions of the behavioral and social sciences. This research, which benefits from cross-national comparisons among the United States, France, Great Britain, and other countries, has been stimulated by an interest in the relations between behavioral and social sciences knowledge and the wider cultural, socioeconomic, and political contexts.

Early studies of the application of behavioral and social sciences knowledge to public policy making proceeded on the model that knowledge use was "instrumental/decisional," in other words, that empirical research findings could be applied to the straightforward solution of well-defined policy problems. Recent empirical research on how political agendas are set and political learning takes place challenges this model. The emerging view stresses the complexity of public policy making and the importance of indirect effects of scientific knowledge on it. Rather than producing engineering-type solutions to narrowly bounded social problems, behavioral and social sciences research more typically provides facts, empirical generalizations, and critical and innovative ideas and perspectives that inform, and sometimes transform, the thinking and, ultimately, the actions of policy makers and the public. Such effects sometimes occur in the near term and sometimes in the long term.

For example, both policy makers and their political supporters and opponents may be more affected by research knowledge as filtered through the media than by data or conclusions directly from research itself. Moreover, the teaching of the behavioral and social sciences to growing cohorts of middle-class American college students may have profoundly affected public policy making throughout the twentieth century by shaping educated people's sense of what social problems could be addressed by governmental action and how they could be addressed. This social science education may also have influenced more general societal phenomena, such as the rise of merit selection as a principle of organizational design and social mobility and the shifting patterns of race relations in the United States.

Current research is focusing particularly on the role of intermediary institutions in bridging the academic and policy worlds. These institutions include governmental units, such as staff offices, agencies, or commissions that draw upon or conduct academic-style research, and nongovernmental organizations, such as philanthropic foundations, think tanks, schools of public policy, and contract research firms. A major contribution could be made by research that traces the full array of interconnections between behavioral and social sciences findings and policy making over considerable periods of time. Such research could carefully follow the concrete processes by which social science knowledge affects the framing of issues and the making of particular policies. It should pay attention to concepts and findings that were potentially available but did not actually influence policy making. Moreover, policies should be traced through their actual implementation and the subsequent assessments of their effects by policy makers and the public.

In addition to questions concerning the application of social and behavioral sciences to policy, other questions surrounding the development and role of these sciences are drawing increased attention. Research is needed on the history of the social science professions, both in the United States and in other countries, with a concentration on the periods in which the major disciplines were founded, the processes that forged new university-based career patterns and new definitions of professional conduct for producers of knowledge about society, and the mobilization of behavioral and social scientists for particular national efforts, such as World War II. Another important topic is how the adoption of nineteenth-century positivist science as a model for the new disciplines—rooted in the profoundly ahistorical sense of "American exceptionalism" that pervaded American culture—generated authority and legitimacy for these professions in the United States. In particular, how did the interests and funding policies of private foundations (especially in the 1920s and 1930s), the policy choices and funding patterns of government (especially since the 1950s), the emergence of think tanks, and the evolution of universities mold the development and carrying out of behavioral and social sciences research agendas.

Most of the studies to date have been conducted in limited settings: one nation or perhaps one agency. More recent work is considering cross-national and comparative-historical dimensions, such as the differences in the structure, function, and uses of social science by governmental commissions in Great Britain, Sweden, and the United States. The international diffusion of prestigious ideas—for example, economic theories such as mercantilism, free trade, Keynesianism, Chicago school monetarism, and various models of modernization, as well as the use of analytic techniques, such as cost-benefit accounting, by the World Bank and other international bodies—has created broad international linkage between intellectuals and social-science-trained "technocrats" in the West and governments in developing nations. Comparative-international studies that track and lead to understanding such intellectual transfers on a global level promise to enrich understanding of national singularities and cross-national regularities in the processes and people that link social science knowledge and policy makers. They can also serve a critical self-reflexive function for behavioral and social scientists, and open the door to more sophisticated future relations between them and other sciences.

INTERNATIONALIZATION

The study of institutional and cultural change within the modernization tradition of research has until recently concentrated on the individual society or nation-state as the basic unit of study. Within this tradition it was widely assumed that the fortunes of different nation-states would be similar during the process of modernization and that each nation's destiny was within its own control. Accordingly, social change was regarded as resulting from the interaction of a number of forces pressing toward modernization (entrepreneurship, generation of savings and capital, labor mobility, enlightened state policy) and a number of opposite forces pressing toward the maintenance of traditional structures (kinship, tribal and communal loyalties, religious beliefs, and loyalties toward localities).

About two decades ago the nation-based focus came under critical reexamination from two sources. The first, largely theoretical in character, came from a number of Latin American and other scholars. Puzzling over certain anomalies in South American history relative to North American history, they argued that the development process in those countries could be better understood not as autonomous and within their control, but as governed—and in some cases deflected or defeated—by the power of external capital emanating mainly from the United States and other developed countries. Parallel theoretical criticism came from scholars who insisted that the long-term structure and short-term pertubations of the worldwide economy dominated the strategies and histories of individual nations, developed or not.

The second source of criticism of the nation-centric approach came from new data from research on contemporary events. The increasing internationalization of the world is a fact that has dominated countless data series and cross-national analyses. There is increasing internationalization of finance and production, largely through the penetration of multinational corporations; labor, as increasing numbers of firms develop the strategy and capability to direct their activities to areas of the world where labor is most economically hired and as laborers migrate more across borders to "informal employment" sectors that are of increasing importance; technology, as nations strive for advantages in the competitive struggle; and culture, as increased interaction among nations spreads new stylistic, political, and business understandings and protocols along with the penetration of television, radio, and written media.

International Finance and Domestic Policy

A particularly active line of research involves the relationship between international finance and the domestic affairs of nations. The increased transnationalization of finance has had contradictory results. While it has clearly limited some of the options of nations, it has made it easier for governments to attract capital for investment purposes or to finance budget deficits. The perceived inability of nations to regulate the Eurocurrency market has meant a partial loss of their control over monetary policy, as has the international debt crisis resulting from loans to developing countries during the 1970s. The breakdown of the Bretton Woods system and the introduction of floating exchange rates has meant greater volatility and increased the impact of international investors on domestic economies. For example, capital flight in France undermined the Mitterand expansion attempt, and the high value of the U.S. dollar in the early 1980s played havoc with U.S. trade. The evidence has led numerous analysts to the view that monetary policy can no longer be pursued nationally but must explicitly take into account policies of all major industrialized countries—an analytical result that has been taken very seriously by the central bankers of those countries.

But this is not to say that countries and their agencies are no longer key actors in international finance. Transnational banks continue to depend on core nation-states to maintain the currencies in which they do business. Floating exchange rates provide somewhat more leeway for differing macroeconomic policies than did the previous fixed exchange rates. Even in developing countries, where the power of the international financial community has grown substantially over the past decade, the effects on state agencies have been paradoxical. Much more than in developed countries, state policies in developing countries have become constrained by the exigencies of international finance. Yet at the same time increased availability of international finance helped place state controlled agencies in a central position vis-à-vis the organization of domestic economies and in certain respects has increased these countries' capacity for autonomous action.

The extractive industries provide a good example of these phenomena. Research has clearly shown that transnational mining corporations are likely to frustrate national development plans by their reluctance to invest their returns in the expansion of local operations, especially when there might be an upward shift in the host country's position in the international division of labor. Yet the same research also shows that countries might diminish their chances of realizing other objectives (for example, access to stable markets) by eliminating such corporate participation in their economies. Other research has shown that it is possible for countries with political will and technocratic competence to induce transnational corporations to conform more closely to certain local goals.

All the research to date reinforces the conclusion that individual countries not only remain important actors in the new internationalized organization of production but also, precisely because of the necessity of bargaining with multinational corporations, have become more rather than less important. Better theoretical models of how state agencies function are needed in order to understand why some countries but not others are able to develop effective institutions for dealing with the new international economic environment. In turn, this success or failure shapes the structure of the international economic system itself.

Cultural and Political Diffusion

Recent research suggests that some developing countries may actively attempt to reorganize their social arrangements around world models as they enter the international system. This possibility challenges theoretical models that have emphasized national uniqueness and the intrasocietal coupling of institutions and models that view the evolution of societal institutions as produced primarily by uniform adaptive responses to the political and economic environment.

Cross-national data on education systems have shown that the expansion, structure, and content of education systems reflect the diffusion of world models as much as the economic or political situation of the country in which they are located. National constitutions have been shown to be remarkably similar in prescribing the rights and duties of citizens, with the similarities becoming greater in the mid-twentieth century. Welfare-state institutions (or, at least, their legal principles) spread rather quickly throughout the entire nation-state system and have evolved in roughly common directions in the most disparate countries. In the economic arena, occupational changes in the contemporary world system have been surprisingly similar; the service sector and its professional component have expanded in parallel ways in countries at different levels of development. Even the responses of states to transnational corporations are derived in large measure from a global culture of bargaining norms. For example, the bargaining position of Papua New Guinea vis-à-vis international copper companies has been found to reflect a global culture created by the accumulated experience of other developing countries as much as its own political history or economic circumstances.

The International Division of Labor

Some developing countries have shifted their position in relation to the international division of labor dramatically. The movement of newly industrialized countries, like Korea and Taiwan—away from the export of traditional, labor-intensive, light manufactured goods toward the export of highly sophisticated high-technology products (such as video cassette recorders)—or like Brazil—away from the export of coffee to steel, automobiles, and airplanes, exemplifies the process. The consequences for trade patterns have been documented, but the institutional dynamics of the process itself have not been well analyzed.

Explanations of the institutional bases of change in the international division of labor are central to theories both of national development and of the international economic system. Fifteen years ago it was assumed that core countries that were "home" to transnational corporations (for example, the United States) would be the beneficiaries of the expansion of those corporations, but subsequent research raised serious questions as to the validity of that assumption. That research contributed to the anxious policy reassessments that have dominated consideration of the new international economy in the United States in the 1980s.

The question of whether the position of the United States in the international division of labor may be slipping has both theoretical and policy aspects. To date, most work on developed and developing countries has been conducted by quite distinct sets of scholars; distinct conceptual frameworks have evolved for studying "development" in contrast to the concept of the political economy of "advanced" industrial countries. In addition, the quality of data available on developing countries is significantly lower than that on developed countries. Scientific norms of methodological rigor tend to discourage researchers from testing their theories beyond the range of cases, usually limited, in which the quality of available data is commensurate with the rigor of the models. Advances in knowledge therefore require sustained and improved access to material and sites, to governments, firms, and banks that are inclined to be secretive on many scores, and in areas of the world in which the political problems facing behavioral and social scholars attempting to do research are particularly delicate.

Past attempts to circumvent these obstacles to comparative research have focused on the support of area studies. The institutional character of area studies programs and centers tended to reward those with a detailed command of specific local knowledge and not to reward those taking approaches that are more broadly theoretical. Allocation of new resources—or modest reallocation of resources currently defined exclusively in terms of geographic or disciplinary categories—into organizational structures explicitly designed to facilitate work that moves across such boundaries would have a major stimulating effect on research on the international division of labor.

Productivity

Another active line of inquiry focuses on the measured slowdown in the growth of productivity, which has been a worldwide phenomenon since about 1970. Productivity growth began to decline somewhat in the late 1960s in the United States and has since been about 2 percent a year less than during the previous 20 years. Europe and Japan have witnessed slowdowns at least as large as in the United States, though starting from higher levels. Particularly large declines in all these countries were noticeable in the immediate aftermath of the 1973 and 1979 oil shocks.

It has proven difficult, however, to determine what has been causing these declines. Neither regulatory interventions, higher energy prices, nor changes in rates of capital formation seem to be direct causes. And the declines were not uniform: productivity increased in agriculture and manufacturing, but not in the service sectors. An influx of "baby boomers" into low-wage service jobs may account for part of the productivity slowdown in the United States, but it would not apply elsewhere.

Several lines of research have emphasized the social dimensions of productivity. One series of important studies finds, for example, that in many settings unionized establishments are more productive than nonunionized ones because of lower quit rates and absenteeism and, perhaps, better treatment of grievances. Other cross-national comparisons suggest that physically comparable plants in Great Britain, Japan, and the United States operate at very different levels of productivity, which also argues for much more intensive consideration of social factors. The very different responses of the American, European, and Japanese economies to the oil shocks have also attracted attention: since 1973 the United States has created more than 20 million new jobs;

Productivity Growth: Output per Hour of Work in Selected Older Industrial Countries Since 1950

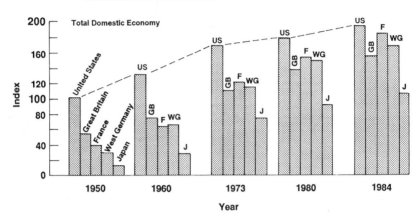

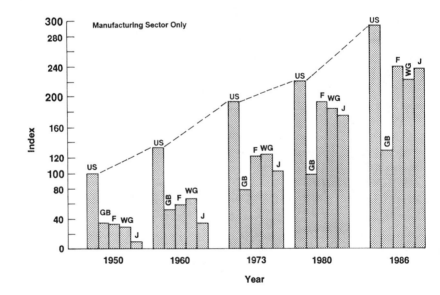

while employment in Europe has been completely stagnant. In Europe it appears that real wages are rigid so that reductions in the standard of living due to oil shocks are transmitted to workers by increased unemployment: that is, when wages rise relative to productivity, unemployment increases. In the United States and perhaps in Japan, it is nominal rather than real wages that are rigid; increases in price levels effectively reduce real wages, which in turn result in maintaining relatively higher levels of employment. The overriding question for research concerns the reasons for the observed differences in the apparent behavior of wages and prices in different countries.

INTERNATIONAL CONFLICT

The four decades since the conclusion of World War II have witnessed a great growth of interest in international politics and security, but the number of active researchers is surprisingly small, partly due to alternating periods of

---

PRODUCTIVITY   What is the nature of international industrial competition? How is productivity related to economic competitiveness? Is the United States in the midst of a major crisis presaging economic decline? The recent history of international economic relations has been volatile, challenging scientists to understand what drives the global movement of products, facilities, and money. This figure displays some important international trends in labor productivity, which is the value of product generated per hour of work, for the five major Western industrial countries.

From the outset and throughout the period since 1950, the U.S. economy has been better at converting labor time into outputs of goods and services than were the economies of other mature industrial nations, particularly in manufacturing. All countries have shown strong gains in this period, especially in manufacturing; the United States still leads in labor productivity, but that lead is now much slimmer.

Overall, the United States remains the most productive economy in terms of output per work hour. The Japanese economy is still relatively low on this measure, at just over half the U.S. level. In contrast, hourly labor productivity in the French economy is high, nearly approaching the U.S. level. However, this performance is less impressive in terms of output per person. The French work a little more than one-half as many hours per capita as the Japanese, due to a shorter work week, more vacations, lower labor-force participation, and higher unemployment. The U.S. population works two-thirds as many hours per capita as the Japanese. The recent great success of Japan in international trade is due not to generalized gains in hourly labor productivity or longer hours, but to other factors, such as greater capital investment, lower real wages, and specific efficiencies in the export-oriented manufacturing sector.

feast and famine in support for researchers. While the world situation dictates a focus on the potential of nuclear war between the United States and the Soviet Union, research also focuses on more general causes of cooperation and conflict. The following kinds of questions are under systematic scrutiny: What is the relationship between national attributes and the domestic and foreign policy of nations? What is the evolving structure of the international political system? What are the causes of international crises and wars? What are the dynamics generally of interaction among nations?

To answer these questions effectively, a systematic research approach is essential. Social and behavioral scientists need, first, to focus on posing the right research questions; second, to develop the right kinds of research designs to address the questions; and, third, to retrieve and generate the most relevant quantitative data bases and qualitative evidence that can generate explanations about them. Above all, the research should be historically and internationally informed.

Superpower Relations

The world has not seen a nuclear exchange and has seen only a handful of major confrontations between the superpowers. Thus, there is no direct evidence bearing on such crucial questions as whether nuclear war can be limited, how decision makers would behave on the brink of war, and the influence of the strategic nuclear balance on the outcomes of confrontations. But there have been 40 years of Soviet-American interaction as superpowers. Despite constraints on access to data that would help illuminate how U.S. and Soviet policies and Soviet-American interaction have evolved, especially at turning points in the Cold War, important studies have been conducted.

For example, some research has focused on the differences between declared American military policy and actual war planning. While the former has often stressed the concept of assured destruction of cities, the latter has always stressed the need to hit a wide variety of military targets. Over time there has been greater consistency in war planning, for which operational requirements and difficulties play a larger role than they do in policy declarations. Researchers have also learned that U.S. presidents are torn between regarding nuclear weapons as extraordinary and divorced from international politics and seeing them as merely very powerful bombs. President Eisenhower, for example, seems to have begun his presidency with the latter perspective and shifted to the former by the time he left office.

In incorporating modern technology into their military establishments, the United States, the Soviet Union, and their respective allies have transformed their military organizations, instituting many surveillance and intelligence operations that were neither necessary nor feasible in earlier times. Many of these activities are conducted daily on a global scale and constitute continuous sources of international tension. Theories of organizational behavior generate a number of empirical questions about both the internal operations of current military establishments and the interactions among them. Those questions concern the type and frequency of operational interactions between military forces, the responsiveness of different levels of the organizational hierarchy, the flexibility of operations, and the differences between normal peacetime and crisis operations.

Decision Making, Beliefs, and Cognitions

Progress that has been made in recent years in understanding decision-making processes (discussed in Chapters 1, 2, and 3) has extended to the study of how national leaders think about national security issues. For example, work on cognition that stresses the importance of beliefs (called scripts or schemata) helps explain both specific intelligence failures (for example, Pearl Harbor, the Iranian revolution) and the general tendency for leaders to be very slow to change their images of other countries. More generally, decision makers, like all people in their daily lives, use shortcuts to making decisions, which conserve their cognitive resources by oversimplifying the world. But this mode of information processing also leads to errors and biases as they use information that is readily available and relatively easy to grasp even if it is not the most relevant for the task at hand. Leaders who are under great pressure to follow a particular course of action—for example, to challenge or try to block the action of another country—are likely to develop an unwarranted belief that the course of action will succeed. The result may be unexpected conflict.

Researchers need to develop a fuller understanding of how policy about national security issues is made. Useful starting points for study include how and why nations perceive others as threats, how images of other states are established and altered, the ways in which conflicts among important values are treated, how statesmen decide that certain threats are so implausible that they can be safely dismissed, and the biases and methods of simplification that characterize adversarial identification and group decision making. The examination of such processes, both within a country and between countries, offers a way of clarifying how and how much international conflict and war can be explained by systemic factors, such as balances of power; by domestic factors, such as specific national capabilities, needs, and demands; by decision-making factors, such as beliefs—including ideologies—and changes in leadership; or by interactions among the three.

One very active research area is the impact of the domestic characteristics of a country on its security-related policies. Internal politics often influences if not dictates, external choices. Historical research into the interwar period, for example, has uncovered the deep disagreements between Great Britain and France on how strong Germany should be permitted to become before being considered a military menace and found that the passivity of political leaders in both countries could be traced to their weakness in the face of conflicting internal demands. At the present time, for example, many security-related issues in the United States—arms control, nuclear deployment, the transfer of technology—are highly partisan political issues. At the same time, of course, international forces shape domestic economic and political life as well. Whether or not a state introduces strict internal political controls, for example, depends in part on the degree to which its security may be threatened by other nations.

Cooperation and Conflict

International politics combines cooperation and conflict, and the analysis of international politics must do the same. The example of the prisoner's dilemma, which incorporates these features, has led to a great deal of work in the field. In its most pristine form, the dilemma is this: in a particular transaction, each person has two options, which can be characterized as to cooperate or not to cooperate. If everyone cooperates, everyone receives a positive but modest return. For each person, however, the temptation not to cooperate is very strong because a noncooperative strategy greatly increases that person's return if most of the other people cooperate. But if no one cooperates, everyone's return is very negative. What is so compellingly perverse about the situation is that, for each individual in each transaction, the noncooperative alternative is at least as good as the cooperative one, regardless of what the others do. So over a series of short-term transactions, self-interest leads everyone to act noncooperatively. But the return from that strategy yields less benefit to every participant than would one of general cooperation.

A formally comparable problem in international politics has been referred to as the security dilemma, in which efforts by one country to maximize its security by arming more heavily have the effect—whether intended or desired—of decreasing the security of other states, which are likely to react by increasing their arms, yielding mutually damaging arms races.

Experimental studies have yielded valuable insights about the ways that modest shifts in payoff influence strategy, the conditions under which the third best outcome occurs, and particularly the conditions under which adversaries are able to and most likely to cooperate with each other. The time perspectives and the relative values of the payoffs to the participants are clearly important: cooperation is most likely when participants expect to have a long series of interactions, not one of which will be decisive; when the gains for exploiting the other and the losses for being exploited are relatively small; and when mutual competition is much worse for both sides than is mutual cooperation. Cooperation is facilitated by contingent strategies such as reciprocity, which is based on the principle of cooperating with another participant when and only when that participant cooperates. Cooperation is also more likely when participants can determine with some certainty whether or not other participants are cooperating, when they are willing and able to reply in kind to the others' behavior, and when the other side realizes this. Under these circumstances, attempting to gain unilateral advantage is less tempting because it is seen as likely to provoke a negative response.

These research findings are promising, especially since a number of alternative hypotheses about the causes of conflict and war have been contradicted by evidence from quantitative studies. Attributes such as a nation's power or its governmental structure have not been shown to be directly related to its involvement in war. Being rich or poor, big or small, and densely or sparsely populated also does not seem to make a country more or less prone to war. And a country's internal political difficulties do not appear to make it more or less likely to engage in conflict. Arms expenditures do tend to be positively related to the incidence of warlike activity, although arms races do not invariably produce wars. The very concept of an "arms race" is undergoing reinterpretation. Arms races have typically been defined as accelerating military expenditures in the face of a potential enemy who is doing likewise, but it has become increasingly clear that the domestic pressure for military expenditures can be a more important factor than imminent international conflict.

Relative power, particularly between bordering nations, has also been shown to be a factor that influences the probability for war. However, contrary to arguments that have been advanced under the theory that a balance of power reduces the likelihood of conflict, significant evidence indicates that wars are most likely between nations with equal rather than unequal power. The common view that wars are typically the consequence of accumulative, escalatory hostile interactions is not supported by analyses of numerous crises and small wars since World War II. Extensive work on the relationship between various structural attributes of the international system—alliance configurations, polarization, power distributions, and status ordering—show that these attributes do affect the level of conflict between the nations in the system, but the patterns are complex.

Because research on international and national security issues is so important for public policy, it is now receiving major funding from private foundations. But most of this funding is aimed at bringing research perspectives to bear on current security policy questions. While these objectives are important, there is a real danger that basic research on cooperation and conflict is being neglected. Furthermore, funding for the development of quantitative data and documentary resources has been sporadic at best. The data sets that do exist are largely the work of a few individual researchers with no guarantee that they will continue to be updated and no clear opportunity for extending and developing the compilations in response to the evolving needs of the research community. Although data collectors are generally aware of each others' material, there is no mechanism to integrate and compare their results. Much of the relevant information that has been produced or gathered by the U.S. government is classified and not readily accessible to scholars. Although many documents are so sensitive that they should remain secret, many are not. For declassified information, a system of coordination and sharing of information is required.

Support for other approaches is also needed. Documents themselves rarely tell the whole story and need to be supplemented by structured interviews, especially with the lower-level officials who have played crucial roles in such areas as American war planning and the analysis of Soviet military posture—and, of course, their Soviet counterparts to the extent possible. More extensive declassification and structured interviewing is in the interest of the government as well as the research community. It would generate research that civil servants, even intelligence specialists, do not have the time or skills to conduct. The greater understanding of current problems that results from careful analysis of earlier situations would benefit the government as well as the research community. For this work, it is especially important to develop efficient means of communication (preprint series, electronic mail and bulletin boards, teleconference systems) and to make arrangements for extended interchange among people who are working on similar problems.

OPPORTUNITIES AND NEEDS

Research on institutions and cultures, like most of the research areas covered in the previous three chapters, calls for a diversity of theories, techniques, and data collections. The study of fertility and migration, for example, uses sophisticated models of decision making, complex statistical analysis of demographic time series, ethnographic study of individuals, families, and communities, and archival investigations. Despite this diversity, however, nearly all areas of research on cultures and institutions, including the areas cited in this chapter, call for historical and cross-national studies: the evolution of human characteristics, changes in family structure and in the major world religions, science and technological competition, nations and transnational corporations, and superpower conflict and cooperation.

Federal and foundation support for historical and cross-national research has been particularly sparse for the past two decades. The Ford Foundation's massive support in the late 1950s and early 1960s was both unprecedented and helpful, but it was short lived; congressional passage of the International Education Act of 1966 was promising, but funds were never provided. There have been no new major initiatives since then. We believe it is time to bring support for this type of research fully back into the research picture, both to build on the innovations made in recent years and to develop new capacities to answer the increasing number of complex questions being raised about institutions and cultures. Overall, we recommend new annual expenditures of $51 million for strengthening research on institutions and cultures.

The biggest need in these areas is for an expansion of support for investigator-initiated grants. As we have discussed above, such grants have been among the most fruitful mechanisms for research progress throughout the behavioral and social sciences; however, there is a particular additional need in the fields discussed in this chapter. Many of the topics discussed above could now take advantage of a major expansion of collaborative work. In the study of inter-nationalization processes, for example, it is very important for small groups of scholars to work together for continuous periods of 1 year or more or to meet periodically for several weeks at a time during 2 or 3 years. In the area of international security, these mechanisms, in conjunction with expanded methods of rapid communication, are particularly important to make better use of quantitative and qualitative data resources and to coordinate research efforts. In the study of science and technology, a fundamental opportunity is emerging to develop the comparative study of public and private institutions through team efforts involving senior and postdoctoral researchers and graduate students.

There are tangible costs associated with the augmentation of these kinds of collaborative research. Not only must more than one researcher be supported, but increased travel is also needed to coordinate research efforts, an expense that is even greater when the collaboration is international. (We note that increased support for international collaborative research may make it necessary for federal funding agencies to reconsider and perhaps change existing rules that hinder or preclude extensive collaborative arrangements with foreign research agencies.) We recommend that, at a minimum, a total of $13 million be added to annual expenditures for investigator-initiated grants on institutions and culture, and that at least $3 million of that increase be directed specifically toward expansion of collaborative research.

There is a substantial need for more graduate and postdoctoral fellowship support to restore the flow of new and talented young researchers into these fields. This requirement is especially marked at the postdoctoral level. We recommend that $6 million be added to annual support of postdoctoral fellowships for research on institutions and cultures and that $2 million be added to support of graduate students.

There are unusually strong opportunities for research workshops and advanced training institutes to be effective in ensuring collaboration, dissemination of developing knowledge, and the use of new techniques in research on institutions and culture. A particularly important opportunity is to train more researchers in the most effective methods of retrieving, coding, analyzing, and synthesizing widely different kinds of information from historical archives and less systematic records and artifacts; this kind of training is often provided to historians during their graduate work, but it is seldom part of the training given to other kinds of scientists. Institutes and workshops are the best mechanisms to develop and upgrade such skills and combine them with other research approaches. We recommend an additional $3 million yearly for research workshops among researchers working on related problems, and $1 million for advanced training institutes.

Equipment needed for research on institutions and culture consists mainly of computer hardware and the associated development of software for individual and group investigators. Two special equipment needs are in archaeological studies, for which dating and other techniques have become complex and demanding, and in demographic studies, for which expanded technological capabilities for data-base management, analysis, and dissemination are needed. We recommend additional annual support of $6 million for equipment, $5 million of which should be for computer-related expenditures.

Studies of institutions and cultures very often require the creation and assembly of diverse kinds of empirical information: economic and other time-series data; survey and other interview data; institutional and cultural products, such as legal codes and documents revealing religious beliefs and practices; and documentary, calendrical, and quantitative materials relating to historically important events such as assassinations, political successions, and wars. This information often presents problems of access.

A similar problem arises in the study of how human institutions and populations evolved: a full palette of continuously improving methods, from radiochemical assays to anatomical imaging and reconstruction to symbolic interpretation are used, but overriding all is the need for sustained access to the geographical locales where rich archaeologically accessible traces of the past and relevant ethological and ecological comparison sites can be systematically probed. First-hand geographic access is also needed for the study of fertility and migrational changes in the developing countries of Africa, the interaction between political change and religious movements in Latin America, and the development of state-based institutions for managing relations with transnational corporate enterprises and developing international market strategies in eastern Asia. Continuity of contact, including the ability to host and visit across national and regional borders on a regular basis, and a sharing of contacts and expertise among researchers, are instrumental to sustaining geographical access, and these capabilities require explicit underwriting. Geographic access also includes photography from aircraft or satellite platforms for archaeological and geographic canvasses.

Many facets of these efforts are in the diplomatic arena, involving agreements between the U.S. government and other governments, American universities and foreign governments, American universities and foreign universities, and individual scientists and other governments. These diplomatic activities are a critical adjunct to the increases in research support that we recommend for grants, fellowships, and workshops to advance cross-national and historical research on institutions and cultures.

A different kind of access problem involves privileged information that has been unavailable to researchers because of respondent privacy or anonymity, national security, or trade secrecy. Classified data of the U.S. government, acquired at great expense for primarily military or national intelligence purposes, could significantly contribute to substantial advances in the study of international security and conflict. Procedures of declassification do not now take into specific account the analytical benefit that may accrue as a result of opening data resources to research scholars. A joint inquiry by the relevant agencies and researchers should consider how to build such considerations into an active declassification procedure. Costs involved at the early stages of this process would be mainly administrative, but subsequent efforts to process previously classified information for research would be significant. We recommend allocating $3 million annually to this effort.

A related kind of data-oriented opportunity is expanded research access to microdata files held by government agencies, such as the Internal Revenue Service, the Social Security Administration, and the Bureau of the Census, and to university-based research files, such as the Panel Study of Income Dynamics. There are substantial research programs that depend on research access to certain portions of the accumulated data (and we have recommended expansion of such programs; see Chapter 3). However, relatively precise locational information about individual respondents over time, which is critical for the study of migration and a variety of regional socioeconomic questions, is generally suppressed or withheld in order to ensure the anonymity of respondents. We recommend a research program to make this locational dimension usable for social science research, and we estimate that $1 million annually is needed for this effort.

To facilitate high-quality comparative research on demographic behavior, it would also be very useful to develop a central-library-type facility in the United States, with demographic data bases available on computer tapes. Some demographic data files are available at only one institution and are largely unknown to researchers located elsewhere, at the same time that many data sets are duplicated in a dozen or more American institutions. Some centralization of data is desirable; such a center could also develop into a disseminator of software and technical expertise on a wide variety of computer-related subjects in population studies. This is basically a matter of coordination of existing resources and may be achieved with relatively little net cost.

A final data-access opportunity concerns the historical records of corporations regarding their proprietary research and development activities, which comprise a resource of great importance to understanding the development of modern science and technology. The precedent established by several corporations in enabling these records to be treated as archives and used for research under professional auspices is a very encouraging initiative. An expanded program of record retention, sampling, codification, and structured interviewing to supplement the archival record is highly desirable. At early stages the cost for the release of these kinds of information would be mainly administrative, but we recommend that a regular extension of this process, including translation of the relevant data into forms usable for research, be undertaken at the level of at least $1 million per year.

Another data resource is the longitudinal, international, quantitative data file developed explicitly for cross-national studies of industrial, commercial, or public sector productivity, international security relations, trends in family and population structure, shifts in religious participation, or the growth of scientific and technological capabilities. These files can be developed from a variety of primary and secondary data sources and linked in various ways depending on the nature of the research questions posed to them. The primary issues concerning these data are stability of support, quality control, standardization across different sources, completeness across periods and parts of the globe, and sheer practical experience in using the data by a large enough set of researchers applying the relevant technical and conceptual tools.

Time-series data over long periods are particularly difficult to find or reconstruct. Although the need for such data was first recognized by scientists in the nineteenth century, and there have been sustained efforts over the last 40 years by the United Nations and other organizations, it has been only partly met. A particular need is to strive for improvement of standardized historical series in censuses, surveys, and medical and educational records. Support for data collection, documentation, and dissemination are essential to ensure that a strong factual base is available for the next steps in comparative research on global processes. We recommend that $8 million be allocated annually to develop, order, and analyze such data sets.

The final category of new research opportunities is the creation of research centers. International scientific centers are a major avenue for stabilizing access to overseas sites and enabling the continuous testing and upgrading of theories and methods in cross-national, historical, and longitudinal research. A new international center with a strong demographic emphasis, to complement those in Bangladesh and Guatemala, would be a valuable spur to research. We recommend support for the planning of such a center, with a requirement for detailed proposals as the basis for full-scale evaluation. The study of modern science and technology, including the behavioral and social sciences and their application, is also ripe for the development of one or more research centers. These kinds of initiatives can involve substantial costs. The establishment and the assumption of basic operating expenses for an international center of the type envisioned could run into several millions of dollars per year, though costs can be shared with other nations. Taking the various possibilities into account, we recommend an annual investment of $7 million in new research centers devoted to the study of cultures and institutions.

5

Methods of Data Collection, Representation, and Analysis

5

Methods of Data Collection, Representation, and Analysis

This chapter concerns research on collecting, representing, and analyzing the data that underlie behavioral and social sciences knowledge. Such research, methodological in character, includes ethnographic and historical approaches, scaling, axiomatic measurement, and statistics, with its important relatives, econometrics and psychometrics. The field can be described as including the self-conscious study of how scientists draw inferences and reach conclusions from observations. Since statistics is the largest and most prominent of methodological approaches and is used by researchers in virtually every discipline, statistical work draws the lion's share of this chapter's attention.

Problems of interpreting data arise whenever inherent variation or measurement fluctuations create challenges to understand data or to judge whether observed relationships are significant, durable, or general. Some examples: Is a sharp monthly (or yearly) increase in the rate of juvenile delinquency (or unemployment) in a particular area a matter for alarm, an ordinary periodic or random fluctuation, or the result of a change or quirk in reporting method? Do the temporal patterns seen in such repeated observations reflect a direct causal mechanism, a complex of indirect ones, or just imperfections in the data? Is a decrease in auto injuries an effect of a new seat-belt law? Are the disagreements among people describing some aspect of a subculture too great to draw valid inferences about that aspect of the culture?

Such issues of inference are often closely connected to substantive theory and specific data, and to some extent it is difficult and perhaps misleading to treat methods of data collection, representation, and analysis separately. This report does so, as do all sciences to some extent, because the methods developed often are far more general than the specific problems that originally gave rise to them. There is much transfer of new ideas from one substantive field to another—and to and from fields outside the behavioral and social sciences. Some of the classical methods of statistics arose in studies of astronomical observations, biological variability, and human diversity. The major growth of the classical methods occurred in the twentieth century, greatly stimulated by problems in agriculture and genetics. Some methods for uncovering geometric structures in data, such as multidimensional scaling and factor analysis, originated in research on psychological problems, but have been applied in many other sciences. Some time-series methods were developed originally to deal with economic data, but they are equally applicable to many other kinds of data.

Within the behavioral and social sciences, statistical methods have been developed in and have contributed to an enormous variety of research, including:

- *In economics:* large-scale models of the U.S. economy; effects of taxation, money supply, and other government fiscal and monetary policies; theories of duopoly, oligopoly, and rational expectations; economic effects of slavery.

- *In psychology:* test calibration; the formation of subjective probabilities, their revision in the light of new information, and their use in decision making; psychiatric epidemiology and mental health program evaluation.

- *In sociology and other fields:* victimization and crime rates; effects of incarceration and sentencing policies; deployment of police and fire-fighting forces; discrimination, antitrust, and regulatory court cases; social networks; population growth and forecasting; and voting behavior.

Even such an abridged listing makes clear that improvements in methodology are valuable across the spectrum of empirical research in the behavioral and social sciences as well as in application to policy questions. Clearly, methodological research serves many different purposes, and there is a need to develop different approaches to serve those different purposes, including exploratory data analysis, scientific inference about hypotheses and population parameters, individual decision making, forecasting what will happen in the event or absence of intervention, and assessing causality from both randomized experiments and observational data.

This discussion of methodological research is divided into three areas: design, representation, and analysis. The efficient design of investigations must take place before data are collected because it involves how much, what kind of, and how data are to be collected. What type of study is feasible: experimental, sample survey, field observation, or other? What variables should be measured, controlled, and randomized? How extensive a subject pool or observational period is appropriate? How can study resources be allocated most effectively among various sites, instruments, and subsamples?

The construction of useful representations of the data involves deciding what kind of formal structure best expresses the underlying qualitative and quantitative concepts that are being used in a given study. For example, cost of living is a simple concept to quantify if it applies to a single individual with unchanging tastes in stable markets (that is, markets offering the same array of goods from year to year at varying prices), but as a national aggregate for millions of households and constantly changing consumer product markets, the cost of living is not easy to specify clearly or measure reliably. Statisticians, economists, sociologists, and other experts have long struggled to make the cost of living a precise yet practicable concept that is also efficient to measure, and they must continually modify it to reflect changing circumstances.

Data analysis covers the final step of characterizing and interpreting research findings: Can estimates of the relations between variables be made? Can some conclusion be drawn about correlation, cause and effect, or trends over time? How uncertain are the estimates and conclusions and can that uncertainty be reduced by analyzing the data in a different way? Can computers be used to display complex results graphically for quicker or better understanding or to suggest different ways of proceeding?

Advances in analysis, data representation, and research design feed into and reinforce one another in the course of actual scientific work. The intersections between methodological improvements and empirical advances are an important aspect of the multidisciplinary thrust of progress in the behavioral and social sciences.

DESIGNS FOR DATA COLLECTION

Four broad kinds of research designs are used in the behavioral and social sciences: experimental, survey, comparative, and ethnographic.

Experimental designs, in either the laboratory or field settings, systematically manipulate a few variables while others that may affect the outcome are held constant, randomized, or otherwise controlled. The purpose of randomized experiments is to ensure that only one or a few variables can systematically affect the results, so that causes can be attributed. Survey designs include the collection and analysis of data from censuses, sample surveys, and longitudinal studies and the examination of various relationships among the observed phenomena. Randomization plays a different role here than in experimental designs: it is used to select members of a sample so that the sample is as representative of the whole population as possible. Comparative designs involve the retrieval of evidence that is recorded in the flow of current or past events in different times or places and the interpretation and analysis of this evidence. Ethnographic designs, also known as participant-observation designs, involve a researcher in intensive and direct contact with a group, community, or population being studied, through participation, observation, and extended interviewing.

Experimental Designs

Laboratory Experiments

Laboratory experiments underlie most of the work reported in Chapter 1, significant parts of Chapter 2, and some of the newest lines of research in Chapter 3. Laboratory experiments extend and adapt classical methods of design first developed, for the most part, in the physical and life sciences and agricultural research. Their main feature is the systematic and independent manipulation of a few variables and the strict control or randomization of all other variables that might affect the phenomenon under study. For example, some studies of animal motivation involve the systematic manipulation of amounts of food and feeding schedules while other factors that may also affect motivation, such as body weight, deprivation, and so on, are held constant. New designs are currently coming into play largely because of new analytic and computational methods (discussed below, in "Advances in Statistical Inference and Analysis").

Two examples of empirically important issues that demonstrate the need for broadening classical experimental approaches are open-ended responses and lack of independence of successive experimental trials. The first concerns the design of research protocols that do not require the strict segregation of the events of an experiment into well-defined trials, but permit a subject to respond at will. These methods are needed when what is of interest is how the respondent chooses to allocate behavior in real time and across continuously available alternatives. Such empirical methods have long been used, but they can generate very subtle and difficult problems in experimental design and subsequent analysis. As theories of allocative behavior of all sorts become more sophisticated and precise, the experimental requirements become more demanding, so the need to better understand and solve this range of design issues is an outstanding challenge to methodological ingenuity.

The second issue arises in repeated-trial designs when the behavior on successive trials, even if it does not exhibit a secular trend (such as a learning curve), is markedly influenced by what has happened in the preceding trial or trials. The more naturalistic the experiment and the more sensitive the measurements taken, the more likely it is that such effects will occur. But such sequential dependencies in observations cause a number of important conceptual and technical problems in summarizing the data and in testing analytical models, which are not yet completely understood. In the absence of clear solutions, such effects are sometimes ignored by investigators, simplifying the data analysis but leaving residues of skepticism about the reliability and significance of the experimental results. With continuing development of sensitive measures in repeated-trial designs, there is a growing need for more advanced concepts and methods for dealing with experimental results that may be influenced by sequential dependencies.

Randomized Field Experiments

The state of the art in randomized field experiments, in which different policies or procedures are tested in controlled trials under real conditions, has advanced dramatically over the past two decades. Problems that were once considered major methodological obstacles—such as implementing randomized field assignment to treatment and control groups and protecting the randomization procedure from corruption—have been largely overcome. While state-of-the-art standards are not achieved in every field experiment, the commitment to reaching them is rising steadily, not only among researchers but also among customer agencies and sponsors.

The health insurance experiment described in Chapter 2 is an example of a major randomized field experiment that has had and will continue to have important policy reverberations in the design of health care financing. Field experiments with the negative income tax (guaranteed minimum income) conducted in the 1970s were significant in policy debates, even before their completion, and provided the most solid evidence available on how tax-based income support programs and marginal tax rates can affect the work incentives and family structures of the poor. Important field experiments have also been carried out on alternative strategies for the prevention of delinquency and other criminal behavior, reform of court procedures, rehabilitative programs in mental health, family planning, and special educational programs, among other areas.

In planning field experiments, much hinges on the definition and design of the experimental cells, the particular combinations needed of treatment and control conditions for each set of demographic or other client sample characteristics, including specification of the minimum number of cases needed in each cell to test for the presence of effects. Considerations of statistical power, client availability, and the theoretical structure of the inquiry enter into such specifications. Current important methodological thresholds are to find better ways of predicting recruitment and attrition patterns in the sample, of designing experiments that will be statistically robust in the face of problematic sample recruitment or excessive attrition, and of ensuring appropriate acquisition and analysis of data on the attrition component of the sample.

Also of major significance are improvements in integrating detailed process and outcome measurements in field experiments. To conduct research on program effects under field conditions requires continual monitoring to determine exactly what is being done—the process—how it corresponds to what was projected at the outset. Relatively unintrusive, inexpensive, and effective implementation measures are of great interest. There is, in parallel, a growing emphasis on designing experiments to evaluate distinct program components in contrast to summary measures of net program effects.

Finally, there is an important opportunity now for further theoretical work to model organizational processes in social settings and to design and select outcome variables that, in the relatively short time of most field experiments, can predict longer-term effects: For example, in job-training programs, what are the effects on the community (role models, morale, referral networks) or on individual skills, motives, or knowledge levels that are likely to translate into sustained changes in career paths and income levels?

Survey Designs

Many people have opinions about how societal mores, economic conditions, and social programs shape lives and encourage or discourage various kinds of behavior. People generalize from their own cases, and from the groups to which they belong, about such matters as how much it costs to raise a child, the extent to which unemployment contributes to divorce, and so on. In fact, however, effects vary so much from one group to another that homespun generalizations are of little use. Fortunately, behavioral and social scientists have been able to bridge the gaps between personal perspectives and collective realities by means of survey research. In particular, governmental information systems include volumes of extremely valuable survey data, and the facility of modern computers to store, disseminate, and analyze such data has significantly improved empirical tests and led to new understandings of social processes.

Within this category of research designs, two major types are distinguished: repeated cross-sectional surveys and longitudinal panel surveys. In addition, and cross-cutting these types, there is a major effort under way to improve and refine the quality of survey data by investigating features of human memory and of question formation that affect survey response.

Repeated cross-sectional designs can either attempt to measure an entire population—as does the oldest U.S. example, the national decennial census—or they can rest on samples drawn from a population. The general principle is to take independent samples at two or more times, measuring the variables of interest, such as income levels, housing plans, or opinions about public affairs, in the same way. The General Social Survey, collected by the National Opinion Research Center with National Science Foundation support, is a repeated cross-sectional data base that was begun in 1972. One methodological question of particular salience in such data is how to adjust for nonresponses and "don't know" responses. Another is how to deal with self-selection bias. For example, to compare the earnings of women and men in the labor force, it would be mistaken to first assume that the two samples of labor-force participants are randomly selected from the larger populations of men and women; instead, one has to consider and incorporate in the analysis the factors that determine who is in the labor force.

In longitudinal panels, a sample is drawn at one point in time and the relevant variables are measured at this and subsequent times for the same people. In more complex versions, some fraction of each panel may be replaced or added to periodically, such as expanding the sample to include households formed by the children of the original sample. An example of panel data developed in this way is the Panel Study of Income Dynamics (PSID), conducted by the University of Michigan since 1968 (discussed in Chapter 3).

Comparing the fertility or income of different people in different circumstances at the same time to find correlations always leaves a large proportion of the variability unexplained, but common sense suggests that much of the unexplained variability is actually explicable. There are systematic reasons for individual outcomes in each person's past achievements, in parental models, upbringing, and earlier sequences of experiences. Unfortunately, asking people about the past is not particularly helpful: people remake their views of the past to rationalize the present and so retrospective data are often of uncertain validity. In contrast, generation-long longitudinal data allow readings on the sequence of past circumstances uncolored by later outcomes. Such data are uniquely useful for studying the causes and consequences of naturally occurring decisions and transitions. Thus, as longitudinal studies continue, quantitative analysis is becoming feasible about such questions as: How are the decisions of individuals affected by parental experience? Which aspects of early decisions constrain later opportunities? And how does detailed background experience leave its imprint? Studies like the two-decade-long PSID are bringing within grasp a complete generational cycle of detailed data on fertility, work life, household structure, and income.

Advances in Longitudinal Designs

Large-scale longitudinal data collection projects are uniquely valuable as vehicles for testing and improving survey research methodology. In ways that lie beyond the scope of a cross-sectional survey, longitudinal studies can sometimes be designed—without significant detriment to their substantive interests—to facilitate the evaluation and upgrading of data quality; the analysis of relative costs and effectiveness of alternative techniques of inquiry; and the standardization or coordination of solutions to problems of method, concept, and measurement across different research domains.

Some areas of methodological improvement include discoveries about the impact of interview mode on response (mail, telephone, face-to-face); the effects of nonresponse on the representativeness of a sample (due to respondents' refusal or interviewers' failure to contact); the effects on behavior of continued participation over time in a sample survey; the value of alternative methods of adjusting for nonresponse and incomplete observations (such as imputation of missing data, variable case weighting); the impact on response of specifying different recall periods, varying the intervals between interviews, or changing the length of interviews; and the comparison and calibration of results obtained by longitudinal surveys, randomized field experiments, laboratory studies, one-time surveys, and administrative records.

It should be especially noted that incorporating improvements in methodology and data quality has been and will no doubt continue to be crucial to the growing success of longitudinal studies. Panel designs are intrinsically more vulnerable than other designs to statistical biases due to cumulative item non-response, sample attrition, time-in-sample effects, and error margins in repeated measures, all of which may produce exaggerated estimates of change. Over time, a panel that was initially representative may become much less representative of a population, not only because of attrition in the sample, but also because of changes in immigration patterns, age structure, and the like. Longitudinal studies are also subject to changes in scientific and societal contexts that may create uncontrolled drifts over time in the meaning of nominally stable questions or concepts as well as in the underlying behavior. Also, a natural tendency to expand over time the range of topics and thus the interview lengths, which increases the burdens on respondents, may lead to deterioration of data quality or relevance. Careful methodological research to understand and overcome these problems has been done, and continued work as a component of new longitudinal studies is certain to advance the overall state of the art.

Longitudinal studies are sometimes pressed for evidence they are not designed to produce: for example, in important public policy questions concerning the impact of government programs in such areas as health promotion, disease prevention, or criminal justice. By using research designs that combine field experiments (with randomized assignment to program and control conditions) and longitudinal surveys, one can capitalize on the strongest merits of each: the experimental component provides stronger evidence for casual statements that are critical for evaluating programs and for illuminating some fundamental theories; the longitudinal component helps in the estimation of long-term program effects and their attenuation. Coupling experiments to ongoing longitudinal studies is not often feasible, given the multiple constraints of not disrupting the survey, developing all the complicated arrangements that go into a large-scale field experiment, and having the populations of interest overlap in useful ways. Yet opportunities to join field experiments to surveys are of great importance. Coupled studies can produce vital knowledge about the empirical conditions under which the results of longitudinal surveys turn out to be similar to—or divergent from—those produced by randomized field experiments. A pattern of divergence and similarity has begun to emerge in coupled studies; additional cases are needed to understand why some naturally occurring social processes and longitudinal design features seem to approximate formal random allocation and others do not. The methodological implications of such new knowledge go well beyond program evaluation and survey research. These findings bear directly on the confidence scientists—and others—can have in conclusions from observational studies of complex behavioral and social processes, particularly ones that cannot be controlled or simulated within the confines of a laboratory environment.

Memory and the Framing of Questions

A very important opportunity to improve survey methods lies in the reduction of nonsampling error due to questionnaire context, phrasing of questions, and, generally, the semantic and social-psychological aspects of surveys. Survey data are particularly affected by the fallibility of human memory and the sensitivity of respondents to the framework in which a question is asked. This sensitivity is especially strong for certain types of attitudinal and opinion questions. Efforts are now being made to bring survey specialists into closer contact with researchers working on memory function, knowledge representation, and language in order to uncover and reduce this kind of error.

Memory for events is often inaccurate, biased toward what respondents believe to be true—or should be true—about the world. In many cases in which data are based on recollection, improvements can be achieved by shifting to techniques of structured interviewing and calibrated forms of memory elicitation, such as specifying recent, brief time periods (for example, in the last seven days) within which respondents recall certain types of events with acceptable accuracy.

Experiments on individual decision making show that the way a question is framed predictably alters the responses. Analysts of survey data find that some small changes in the wording of certain kinds of questions can produce large differences in the answers, although other wording changes have little effect. Even simply changing the order in which some questions are presented can produce large differences, although for other questions the order of presentation does not matter. For example, the following questions were among those asked in one wave of the General Social Survey:

- "Taking things altogether, how would you describe your marriage? Would you say that your marriage is very happy, pretty happy, or not too happy?"

- "Taken altogether how would you say things are these days — would you say you are very happy, pretty happy, or not too happy?"

Presenting this sequence in both directions on different forms showed that the order affected answers to the general happiness question but did not change the marital happiness question: responses to the specific issue swayed subsequent responses to the general one, but not vice versa. The explanations for and implications of such order effects on the many kinds of questions and sequences that can be used are not simple matters. Further experimentation on the design of survey instruments promises not only to improve the accuracy and reliability of survey research, but also to advance understanding of how people think about and evaluate their behavior from day to day.

Comparative Designs

Both experiments and surveys involve interventions or questions by the scientist, who then records and analyzes the responses. In contrast, many bodies of social and behavioral data of considerable value are originally derived from records or collections that have accumulated for various nonscientific reasons, quite often administrative in nature, in firms, churches, military organizations, and governments at all levels. Data of this kind can sometimes be subjected to careful scrutiny, summary, and inquiry by historians and social scientists, and statistical methods have increasingly been used to develop and evaluate inferences drawn from such data. Some of the main comparative approaches are cross-national aggregate comparisons, selective comparison of a limited number of cases, and historical case studies.

Among the more striking problems facing the scientist using such data are the vast differences in what has been recorded by different agencies whose behavior is being compared (this is especially true for parallel agencies in different nations), the highly unrepresentative or idiosyncratic sampling that can occur in the collection of such data, and the selective preservation and destruction of records. Means to overcome these problems form a substantial methodological research agenda in comparative research. An example of the method of cross-national aggregative comparisons is found in investigations by political scientists and sociologists of the factors that underlie differences in the vitality of institutions of political democracy in different societies. Some investigators have stressed the existence of a large middle class, others the level of education of a population, and still others the development of systems of mass communication. In cross-national aggregate comparisons, a large number of nations are arrayed according to some measures of political democracy and then attempts are made to ascertain the strength of correlations between these and the other variables. In this line of analysis it is possible to use a variety of statistical cluster and regression techniques to isolate and assess the possible impact of certain variables on the institutions under study. While this kind of research is cross-sectional in character, statements about historical processes are often invoked to explain the correlations.

More limited selective comparisons, applied by many of the classic theorists, involve asking similar kinds of questions but over a smaller range of societies. Why did democracy develop in such different ways in America, France, and England? Why did northeastern Europe develop rational bourgeois capitalism, in contrast to the Mediterranean and Asian nations? Modern scholars have turned their attention to explaining, for example, differences among types of fascism between the two World Wars, and similarities and differences among modern state welfare systems, using these comparisons to unravel the salient causes. The questions asked in these instances are inevitably historical ones.

Historical case studies involve only one nation or region, and so they may not be geographically comparative. However, insofar as they involve tracing the transformation of a society's major institutions and the role of its main shaping events, they involve a comparison of different periods of a nation's or a region's history. The goal of such comparisons is to give a systematic account of the relevant differences. Sometimes, particularly with respect to the ancient societies, the historical record is very sparse, and the methods of history and archaeology mesh in the reconstruction of complex social arrangements and patterns of change on the basis of few fragments.

Like all research designs, comparative ones have distinctive vulnerabilities and advantages: One of the main advantages of using comparative designs is that they greatly expand the range of data, as well as the amount of variation in those data, for study. Consequently, they allow for more encompassing explanations and theories that can relate highly divergent outcomes to one another in the same framework. They also contribute to reducing any cultural biases or tendencies toward parochialism among scientists studying common human phenomena.

One main vulnerability in such designs arises from the problem of achieving comparability. Because comparative study involves studying societies and other units that are dissimilar from one another, the phenomena under study usually occur in very different contexts—so different that in some cases what is called an event in one society cannot really be regarded as the same type of event in another. For example, a vote in a Western democracy is different from a vote in an Eastern bloc country, and a voluntary vote in the United States means something different from a compulsory vote in Australia. These circumstances make for interpretive difficulties in comparing aggregate rates of voter turnout in different countries.

The problem of achieving comparability appears in historical analysis as well. For example, changes in laws and enforcement and recording procedures over time change the definition of what is and what is not a crime, and for that reason it is difficult to compare the crime rates over time. Comparative re-searchers struggle with this problem continually, working to fashion equivalent measures; some have suggested the use of different measures (voting, letters to the editor, street demonstration) in different societies for common variables (political participation), to try to take contextual factors into account and to achieve truer comparability.

A second vulnerability is controlling variation. Traditional experiments make conscious and elaborate efforts to control the variation of some factors and thereby assess the causal significance of others. In surveys as well as experiments, statistical methods are used to control sources of variation and assess suspected causal significance. In comparative and historical designs, this kind of control is often difficult to attain because the sources of variation are many and the number of cases few. Scientists have made efforts to approximate such control in these cases of "many variables, small N." One is the method of paired comparisons. If an investigator isolates 15 American cities in which racial violence has been recurrent in the past 30 years, for example, it is helpful to match them with 15 cities of similar population size, geographical region, and size of minorities—such characteristics are controls—and then search for systematic differences between the two sets of cities. Another method is to select, for comparative purposes, a sample of societies that resemble one another in certain critical ways, such as size, common language, and common level of development, thus attempting to hold these factors roughly constant, and then seeking explanations among other factors in which the sampled societies differ from one another.

Ethnographic Designs

Traditionally identified with anthropology, ethnographic research designs are playing increasingly significant roles in most of the behavioral and social sciences. The core of this methodology is participant-observation, in which a researcher spends an extended period of time with the group under study, ideally mastering the local language, dialect, or special vocabulary, and participating in as many activities of the group as possible. This kind of participant-observation is normally coupled with extensive open-ended interviewing, in which people are asked to explain in depth the rules, norms, practices, and beliefs through which (from their point of view) they conduct their lives. A principal aim of ethnographic study is to discover the premises on which those rules, norms, practices, and beliefs are built.

The use of ethnographic designs by anthropologists has contributed significantly to the building of knowledge about social and cultural variation. And while these designs continue to center on certain long-standing features— extensive face-to-face experience in the community, linguistic competence, participation, and open-ended interviewing—there are newer trends in ethnographic work. One major trend concerns its scale. Ethnographic methods were originally developed largely for studying small-scale groupings known variously as village, folk, primitive, preliterate, or simple societies. Over the decades, these methods have increasingly been applied to the study of small groups and networks within modern (urban, industrial, complex) society, including the contemporary United States. The typical subjects of ethnographic study in modern society are small groups or relatively small social networks, such as outpatient clinics, medical schools, religious cults and churches, ethnically distinctive urban neighborhoods, corporate offices and factories, and government bureaus and legislatures.

As anthropologists moved into the study of modern societies, researchers in other disciplines—particularly sociology, psychology, and political science—began using ethnographic methods to enrich and focus their own insights and findings. At the same time, studies of large-scale structures and processes have been aided by the use of ethnographic methods, since most large-scale changes work their way into the fabric of community, neighborhood, and family, affecting the daily lives of people. Ethnographers have studied, for example, the impact of new industry and new forms of labor in "backward" regions; the impact of state-level birth control policies on ethnic groups; and the impact on residents in a region of building a dam or establishing a nuclear waste dump. Ethnographic methods have also been used to study a number of social processes that lend themselves to its particular techniques of observation and interview—processes such as the formation of class and racial identities, bureaucratic behavior, legislative coalitions and outcomes, and the formation and shifting of consumer tastes.

Advances in structured interviewing (see above) have proven especially powerful in the study of culture. Techniques for understanding kinship systems, concepts of disease, color terminologies, ethnobotany, and ethnozoology have been radically transformed and strengthened by coupling new interviewing methods with modern measurement and scaling techniques (see below). These techniques have made possible more precise comparisons among cultures and identification of the most competent and expert persons within a culture. The next step is to extend these methods to study the ways in which networks of propositions (such as boys like sports, girls like babies) are organized to form belief systems. Much evidence suggests that people typically represent the world around them by means of relatively complex cognitive models that involve interlocking propositions. The techniques of scaling have been used to develop models of how people categorize objects, and they have great potential for further development, to analyze data pertaining to cultural propositions.

Ideological Systems

Perhaps the most fruitful area for the application of ethnographic methods in recent years has been the systematic study of ideologies in modern society. Earlier studies of ideology were in small-scale societies that were rather homogeneous. In these studies researchers could report on a single culture, a uniform system of beliefs and values for the society as a whole. Modern societies are much more diverse both in origins and number of subcultures, related to different regions, communities, occupations, or ethnic groups. Yet these sub-cultures and ideologies share certain underlying assumptions or at least must find some accommodation with the dominant value and belief systems in the society.

The challenge is to incorporate this greater complexity of structure and process into systematic descriptions and interpretations. One line of work carried out by researchers has tried to track the ways in which ideologies are created, transmitted, and shared among large populations that have tradition-ally lacked the social mobility and communications technologies of the West. This work has concentrated on large-scale civilizations such as China, India, and Central America. Gradually, the focus has generalized into a concern with the relationship between the great traditions—the central lines of cosmopolitan Confucian, Hindu, or Mayan culture, including aesthetic standards, irrigation technologies, medical systems, cosmologies and calendars, legal codes, poetic genres, and religious doctrines and rites—and the little traditions, those iden-tified with rural, peasant communities. How are the ideological doctrines and cultural values of the urban elites, the great traditions, transmitted to local communities? How are the little traditions, the ideas from the more isolated, less literate, and politically weaker groups in society, transmitted to the elites?

India and southern Asia have been fruitful areas for ethnographic research on these questions. The great Hindu tradition was present in virtually all local contexts through the presence of high-caste individuals in every community. It operated as a pervasive standard of value for all members of society, even in the face of strong little traditions. The situation is surprisingly akin to that of modern, industrialized societies. The central research questions are the degree and the nature of penetration of dominant ideology, even in groups that appear marginal and subordinate and have no strong interest in sharing the dominant value system. In this connection the lowest and poorest occupational caste—the untouchables—serves as an ultimate test of the power of ideology and cultural beliefs to unify complex hierarchical social systems.

Historical Reconstruction

Another current trend in ethnographic methods is its convergence with archival methods. One joining point is the application of descriptive and in-terpretative procedures used by ethnographers to reconstruct the cultures that created historical documents, diaries, and other records, to interview history, so to speak. For example, a revealing study showed how the Inquisition in the Italian countryside between the 1570s and 1640s gradually worked subtle changes in an ancient fertility cult in peasant communities; the peasant beliefs and rituals assimilated many elements of witchcraft after learning them from their persecutors. A good deal of social history—particularly that of the fam-ily—has drawn on discoveries made in the ethnographic study of primitive societies. As described in Chapter 4, this particular line of inquiry rests on a marriage of ethnographic, archival, and demographic approaches.

Other lines of ethnographic work have focused on the historical dimensions of nonliterate societies. A strikingly successful example in this kind of effort is a study of head-hunting. By combining an interpretation of local oral tradition with the fragmentary observations that were made by outside observers (such as missionaries, traders, colonial officials), historical fluctuations in the rate and significance of head-hunting were shown to be partly in response to such international forces as the great depression and World War II. Researchers are also investigating the ways in which various groups in contemporary societies invent versions of traditions that may or may not reflect the actual history of the group. This process has been observed among elites seeking political and cultural legitimation and among hard-pressed minorities (for example, the Basque in Spain, the Welsh in Great Britain) seeking roots and political mobilization in a larger society.

Ethnography is a powerful method to record, describe, and interpret the system of meanings held by groups and to discover how those meanings affect the lives of group members. It is a method well adapted to the study of situations in which people interact with one another and the researcher can interact with them as well, so that information about meanings can be evoked and observed. Ethnography is especially suited to exploration and elucidation of unsuspected connections; ideally, it is used in combination with other methods—experimental, survey, or comparative—to establish with precision the relative strengths and weaknesses of such connections. By the same token, experimental, survey, and comparative methods frequently yield connections, the meaning of which is unknown; ethnographic methods are a valuable way to determine them.

MODELS FOR REPRESENTING PHENOMENA

The objective of any science is to uncover the structure and dynamics of the phenomena that are its subject, as they are exhibited in the data. Scientists continuously try to describe possible structures and ask whether the data can, with allowance for errors of measurement, be described adequately in terms of them. Over a long time, various families of structures have recurred throughout many fields of science; these structures have become objects of study in their own right, principally by statisticians, other methodological specialists, applied mathematicians, and philosophers of logic and science. Methods have evolved to evaluate the adequacy of particular structures to account for particular types of data. In the interest of clarity we discuss these structures in this section and the analytical methods used for estimation and evaluation of them in the next section, although in practice they are closely intertwined.

A good deal of mathematical and statistical modeling attempts to describe the relations, both structural and dynamic, that hold among variables that are presumed to be representable by numbers. Such models are applicable in the behavioral and social sciences only to the extent that appropriate numerical measurement can be devised for the relevant variables. In many studies the phenomena in question and the raw data obtained are not intrinsically numerical, but qualitative, such as ethnic group identifications. The identifying numbers used to code such questionnaire categories for computers are no more than labels, which could just as well be letters or colors. One key question is whether there is some natural way to move from the qualitative aspects of such data to a structural representation that involves one of the well-understood numerical or geometric models or whether such an attempt would be inherently inappropriate for the data in question. The decision as to whether or not particular empirical data can be represented in particular numerical or more complex structures is seldom simple, and strong intuitive biases or a priori assumptions about what can and cannot be done may be misleading.

Recent decades have seen rapid and extensive development and application of analytical methods attuned to the nature and complexity of social science data. Examples of nonnumerical modeling are increasing. Moreover, the widespread availability of powerful computers is probably leading to a qualitative revolution, it is affecting not only the ability to compute numerical solutions to numerical models, but also to work out the consequences of all sorts of structures that do not involve numbers at all. The following discussion gives some indication of the richness of past progress and of future prospects although it is by necessity far from exhaustive.

In describing some of the areas of new and continuing research, we have organized this section on the basis of whether the representations are fundamentally probabilistic or not. A further useful distinction is between representations of data that are highly discrete or categorical in nature (such as whether a person is male or female) and those that are continuous in nature (such as a person's height). Of course, there are intermediate cases involving both types of variables, such as color stimuli that are characterized by discrete hues (red, green) and a continuous luminance measure. Probabilistic models lead very naturally to questions of estimation and statistical evaluation of the correspondence between data and model. Those that are not probabilistic involve additional problems of dealing with and representing sources of variability that are not explicitly modeled. At the present time, scientists understand some aspects of structure, such as geometries, and some aspects of randomness, as embodied in probability models, but do not yet adequately understand how to put the two together in a single unified model. Table 5-1 outlines the way we have organized this discussion and shows where the examples in this section lie.

Probability Models

Some behavioral and social sciences variables appear to be more or less continuous, for example, utility of goods, loudness of sounds, or risk associated with uncertain alternatives. Many other variables, however, are inherently cat-

TABLE 5-1   A Classification of Structural Models

Nature of the Representation	Nature of the Variables	
	Categorical	Continuous
Probabilistic	Log-linear and related models	Multi-item measurement
	Event histories	Nonlinear, nonadditive models
Geometric and algebraic	Clustering	Scaling
	Network models	Ordered factorial systems

egorical, often with only two or a few values possible: for example, whether a person is in or out of school, employed or not employed, identifies with a major political party or political ideology. And some variables, such as moral attitudes, are typically measured in research with survey questions that allow only categorical responses. Much of the early probability theory was formulated only for continuous variables; its use with categorical variables was not really justified, and in some cases it may have been misleading. Recently, very significant advances have been made in how to deal explicitly with categorical variables. This section first describes several contemporary approaches to models involving categorical variables, followed by ones involving continuous representations.

Log-Linear Models for Categorical Variables

Many recent models for analyzing categorical data of the kind usually displayed as counts (cell frequencies) in multidimensional contingency tables are subsumed under the general heading of log-linear models; that is, linear models in the natural logarithms of the expected counts in each cell in the table. These recently developed forms of statistical analysis allow one to partition variability due to various sources in the distribution of categorical attributes, and to isolate the effects of particular variables or combinations of them.

Present log-linear models were first developed and used by statisticians and sociologists and then found extensive application in other social and behavioral sciences disciplines. When applied, for instance, to the analysis of social mobility, such models separate factors of occupational supply and demand from other factors that impede or propel movement up and down the social hierarchy. With such models, for example, researchers discovered the surprising fact that occupational mobility patterns are strikingly similar in many nations of the world (even among disparate nations like the United States and most of the Eastern European socialist countries), and from one time period to another, once allowance is made for differences in the distributions of occupations. The log-linear and related kinds of models have also made it possible to identify and analyze systematic differences in mobility among nations and across time. As another example of applications, psychologists and others have used log-linear models to analyze attitudes and their determinants and to link attitudes to behavior. These methods have also diffused to and been used extensively in the medical and biological sciences.

Regression Models for Categorical Variables

Models that permit one variable to be explained or predicted by means of others, called regression models, are the workhorses of much applied statistics; this is especially true when the dependent (explained) variable is continuous. For a two-valued dependent variable, such as alive or dead, models and approximate theory and computational methods for one explanatory variable were developed in biometry about 50 years ago. Computer programs able to handle many explanatory variables, continuous or categorical, are readily available today. Even now, however, the accuracy of the approximate theory on given data is an open question.

Using classical utility theory, economists have developed discrete choice models that turn out to be somewhat related to the log-linear and categorical regression models. Models for limited dependent variables, especially those that cannot take on values above or below a certain level (such as weeks unemployed, number of children, and years of schooling) have been used profitably in economics and in some other areas. For example, censored normal variables (called tobits in economics), in which observed values outside certain limits are simply counted, have been used in studying decisions to go on in school. It will require further research and development to incorporate information about limited ranges of variables fully into the main multivariate methodologies. In addition, with respect to the assumptions about distribution and functional form conventionally made in discrete response models, some new methods are now being developed that show promise of yielding reliable inferences without making unrealistic assumptions; further research in this area promises significant progress.

One problem arises from the fact that many of the categorical variables collected by the major data bases are ordered. For example, attitude surveys frequently use a 3-, 5-, or 7-point scale (from high to low) without specifying numerical intervals between levels. Social class and educational levels are often described by ordered categories. Ignoring order information, which many traditional statistical methods do, may be inefficient or inappropriate, but replacing the categories by successive integers or other arbitrary scores may distort the results. (For additional approaches to this question, see sections below on ordered structures.) Regression-like analysis of ordinal categorical variables is quite well developed, but their multivariate analysis needs further research. New log-bilinear models have been proposed, but to date they deal specifically with only two or three categorical variables. Additional research extending the new models, improving computational algorithms, and integrating the models with work on scaling promise to lead to valuable new knowledge.

Models for Event Histories

Event-history studies yield the sequence of events that respondents to a survey sample experience over a period of time; for example, the timing of marriage, childbearing, or labor force participation. Event-history data can be used to study educational progress, demographic processes (migration, fertility, and mortality), mergers of firms, labor market behavior, and even riots, strikes, and revolutions. As interest in such data has grown, many researchers have turned to models that pertain to changes in probabilities over time to describe when and how individuals move among a set of qualitative states.

Much of the progress in models for event-history data builds on recent developments in statistics and biostatistics for life-time, failure-time, and hazard models. Such models permit the analysis of qualitative transitions in a population whose members are undergoing partially random organic deterioration, mechanical wear, or other risks over time. With the increased complexity of event-history data that are now being collected, and the extension of event-history data bases over very long periods of time, new problems arise that cannot be effectively handled by older types of analysis. Among the problems are repeated transitions, such as between unemployment and employment or marriage and divorce; more than one time variable (such as biological age, calendar time, duration in a stage, and time exposed to some specified condition); latent variables (variables that are explicitly modeled even though not observed); gaps in the data; sample attrition that is not randomly distributed over the categories; and respondent difficulties in recalling the exact timing of events.

Models for Multiple-Item Measurement

For a variety of reasons, researchers typically use multiple measures (or multiple indicators) to represent theoretical concepts. Sociologists, for example, often rely on two or more variables (such as occupation and education) to measure an individual's socioeconomic position; educational psychologists ordinarily measure a student's ability with multiple test items. Despite the fact that the basic observations are categorical, in a number of applications this is interpreted as a partitioning of something continuous. For example, in test theory one thinks of the measures of both item difficulty and respondent ability as continuous variables, possibly multidimensional in character.

Classical test theory and newer item-response theories in psychometrics deal with the extraction of information from multiple measures. Testing, which is a major source of data in education and other areas, results in millions of test items stored in archives each year for purposes ranging from college admissions to job-training programs for industry. One goal of research on such test data is to be able to make comparisons among persons or groups even when different test items are used. Although the information collected from each respondent is intentionally incomplete in order to keep the tests short and simple, item-response techniques permit researchers to reconstitute the fragments into an accurate picture of overall group proficiencies. These new methods provide a better theoretical handle on individual differences, and they are expected to be extremely important in developing and using tests. For example, they have been used in attempts to equate different forms of a test given in successive waves during a year, a procedure made necessary in large-scale testing programs by legislation requiring disclosure of test-scoring keys at the time results are given.

An example of the use of item-response theory in a significant research effort is the National Assessment of Educational Progress (NAEP). The goal of this project is to provide accurate, nationally representative information on the average (rather than individual) proficiency of American children in a wide variety of academic subjects as they progress through elementary and secondary school. This approach is an improvement over the use of trend data on university entrance exams, because NAEP estimates of academic achievements (by broad characteristics such as age, grade, region, ethnic background, and so on) are not distorted by the self-selected character of those students who seek admission to college, graduate, and professional programs.

Item-response theory also forms the basis of many new psychometric instruments, known as computerized adaptive testing, currently being implemented by the U.S. military services and under additional development in many testing organizations. In adaptive tests, a computer program selects items for each examinee based upon the examinee's success with previous items. Generally, each person gets a slightly different set of items and the equivalence of scale scores is established by using item-response theory. Adaptive testing can greatly reduce the number of items needed to achieve a given level of measurement accuracy.

Nonlinear, Nonadditive Models

Virtually all statistical models now in use impose a linearity or additivity assumption of some kind, sometimes after a nonlinear transformation of variables. Imposing these forms on relationships that do not, in fact, possess them may well result in false descriptions and spurious effects. Unwary users, especially of computer software packages, can easily be misled. But more realistic nonlinear and nonadditive multivariate models are becoming available. Extensive use with empirical data is likely to force many changes and enhancements in such models and stimulate quite different approaches to nonlinear multivariate analysis in the next decade.

Geometric and Algebraic Models

Geometric and algebraic models attempt to describe underlying structural relations among variables. In some cases they are part of a probabilistic approach, such as the algebraic models underlying regression or the geometric representations of correlations between items in a technique called factor analysis. In other cases, geometric and algebraic models are developed without explicitly modeling the element of randomness or uncertainty that is always present in the data. Although this latter approach to behavioral and social sciences problems has been less researched than the probabilistic one, there are some advantages in developing the structural aspects independent of the statistical ones. We begin the discussion with some inherently geometric representations and then turn to numerical representations for ordered data.

Although geometry is a huge mathematical topic, little of it seems directly applicable to the kinds of data encountered in the behavioral and social sciences. A major reason is that the primitive concepts normally used in geometry—points, lines, coincidence—do not correspond naturally to the kinds of qualitative observations usually obtained in behavioral and social sciences contexts. Nevertheless, since geometric representations are used to reduce bodies of data, there is a real need to develop a deeper understanding of when such representations of social or psychological data make sense. Moreover, there is a practical need to understand why geometric computer algorithms, such as those of multidimensional scaling, work as well as they apparently do. A better understanding of the algorithms will increase the efficiency and appropriateness of their use, which becomes increasingly important with the widespread availability of scaling programs for microcomputers.

Scaling

Over the past 50 years several kinds of well-understood scaling techniques have been developed and widely used to assist in the search for appropriate geometric representations of empirical data. The whole field of scaling is now entering a critical juncture in terms of unifying and synthesizing what earlier appeared to be disparate contributions. Within the past few years it has become apparent that several major methods of analysis, including some that are based on probabilistic assumptions, can be unified under the rubric of a single generalized mathematical structure. For example, it has recently been demonstrated that such diverse approaches as nonmetric multidimensional scaling, principal-components analysis, factor analysis, correspondence analysis, and log-linear analysis have more in common in terms of underlying mathematical structure than had earlier been realized.

Nonmetric multidimensional scaling is a method that begins with data about the ordering established by subjective similarity (or nearness) between pairs of stimuli. The idea is to embed the stimuli into a metric space (that is, a geometry with a measure of distance between points) in such a way that distances between points corresponding to stimuli exhibit the same ordering as do the data. This method has been successfully applied to phenomena that, on other grounds, are known to be describable in terms of a specific geometric structure; such applications were used to validate the procedures. Such validation was done, for example, with respect to the perception of colors, which are known to be describable in terms of a particular three-dimensional structure known as the Euclidean color coordinates. Similar applications have been made with Morse code symbols and spoken phonemes. The technique is now used in some biological and engineering applications, as well as in some of the social sciences, as a method of data exploration and simplification.

One question of interest is how to develop an axiomatic basis for various geometries using as a primitive concept an observable such as the subject's ordering of the relative similarity of one pair of stimuli to another, which is the typical starting point of such scaling. The general task is to discover properties of the qualitative data sufficient to ensure that a mapping into the geometric structure exists and, ideally, to discover an algorithm for finding it. Some work of this general type has been carried out: for example, there is an elegant set of axioms based on laws of color matching that yields the three-dimensional vectorial representation of color space. But the more general problem of understanding the conditions under which the multidimensional scaling algorithms are suitable remains unsolved. In addition, work is needed on understanding more general, non-Euclidean spatial models.

Ordered Factorial Systems

One type of structure common throughout the sciences arises when an ordered dependent variable is affected by two or more ordered independent variables. This is the situation to which regression and analysis-of-variance models are often applied; it is also the structure underlying the familiar physical identities, in which physical units are expressed as products of the powers of other units (for example, energy has the unit of mass times the square of the unit of distance divided by the square of the unit of time).

There are many examples of these types of structures in the behavioral and social sciences. One example is the ordering of preference of commodity bundles—collections of various amounts of commodities—which may be revealed directly by expressions of preference or indirectly by choices among alternative sets of bundles. A related example is preferences among alternative courses of action that involve various outcomes with differing degrees of uncertainty; this is one of the more thoroughly investigated problems because of its potential importance in decision making. A psychological example is the trade-off between delay and amount of reward, yielding those combinations that are equally reinforcing. In a common, applied kind of problem, a subject is given descriptions of people in terms of several factors, for example, intelligence, creativity, diligence, and honesty, and is asked to rate them according to a criterion such as suitability for a particular job.

In all these cases and a myriad of others like them the question is whether the regularities of the data permit a numerical representation. Initially, three types of representations were studied quite fully: the dependent variable as a sum, a product, or a weighted average of the measures associated with the independent variables. The first two representations underlie some psychological and economic investigations, as well as a considerable portion of physical measurement and modeling in classical statistics. The third representation, averaging, has proved most useful in understanding preferences among uncertain outcomes and the amalgamation of verbally described traits, as well as some physical variables.

For each of these three cases—adding, multiplying, and averaging—researchers know what properties or axioms of order the data must satisfy for such a numerical representation to be appropriate. On the assumption that one or another of these representations exists, and using numerical ratings by subjects instead of ordering, a scaling technique called functional measurement (referring to the function that describes how the dependent variable relates to the independent ones) has been developed and applied in a number of domains. What remains problematic is how to encompass at the ordinal level the fact that some random error intrudes into nearly all observations and then to show how that randomness is represented at the numerical level; this continues to be an unresolved and challenging research issue.

During the past few years considerable progress has been made in understanding certain representations inherently different from those just discussed. The work has involved three related thrusts. The first is a scheme of classifying structures according to how uniquely their representation is constrained. The three classical numerical representations are known as ordinal, interval, and ratio scale types. For systems with continuous numerical representations and of scale type at least as rich as the ratio one, it has been shown that only one additional type can exist. A second thrust is to accept structural assumptions, like factorial ones, and to derive for each scale the possible functional relations among the independent variables. And the third thrust is to develop axioms for the properties of an order relation that leads to the possible representations. Much is now known about the possible nonadditive representations of both the multifactor case and the one where stimuli can be combined, such as combining sound intensities.

Closely related to this classification of structures is the question: What statements, formulated in terms of the measures arising in such representations, can be viewed as meaningful in the sense of corresponding to something empirical? Statements here refer to any scientific assertions, including statistical ones, formulated in terms of the measures of the variables and logical and mathematical connectives. These are statements for which asserting truth or falsity makes sense. In particular, statements that remain invariant under certain symmetries of structure have played an important role in classical geometry, dimensional analysis in physics, and in relating measurement and statistical models applied to the same phenomenon. In addition, these ideas have been used to construct models in more formally developed areas of the behavioral and social sciences, such as psychophysics. Current research has emphasized the communality of these historically independent developments and is attempting both to uncover systematic, philosophically sound arguments as to why invariance under symmetries is as important as it appears to be and to understand what to do when structures lack symmetry, as, for example, when variables have an inherent upper bound.

Clustering

Many subjects do not seem to be correctly represented in terms of distances in continuous geometric space. Rather, in some cases, such as the relations among meanings of words—which is of great interest in the study of memory representations—a description in terms of tree-like, hierarchial structures appears to be more illuminating. This kind of description appears appropriate both because of the categorical nature of the judgments and the hierarchial, rather than trade-off, nature of the structure. Individual items are represented as the terminal nodes of the tree, and groupings by different degrees of similarity are shown as intermediate nodes, with the more general groupings occurring nearer the root of the tree. Clustering techniques, requiring considerable computational power, have been and are being developed. Some successful applications exist, but much more refinement is anticipated.

Network Models

Several other lines of advanced modeling have progressed in recent years, opening new possibilities for empirical specification and testing of a variety of theories. In social network data, relationships among units, rather than the units themselves, are the primary objects of study: friendships among persons, trade ties among nations, cocitation clusters among research scientists, interlocking among corporate boards of directors. Special models for social network data have been developed in the past decade, and they give, among other things, precise new measures of the strengths of relational ties among units. A major challenge in social network data at present is to handle the statistical dependence that arises when the units sampled are related in complex ways.

STATISTICAL INFERENCE AND ANALYSIS

As was noted earlier, questions of design, representation, and analysis are intimately intertwined. Some issues of inference and analysis have been discussed above as related to specific data collection and modeling approaches. This section discusses some more general issues of statistical inference and advances in several current approaches to them.

Causal Inference

Behavioral and social scientists use statistical methods primarily to infer the effects of treatments, interventions, or policy factors. Previous chapters included many instances of causal knowledge gained this way. As noted above, the large experimental study of alternative health care financing discussed in Chapter 2 relied heavily on statistical principles and techniques, including randomization, in the design of the experiment and the analysis of the resulting data. Sophisticated designs were necessary in order to answer a variety of questions in a single large study without confusing the effects of one program difference (such as prepayment or fee for service) with the effects of another (such as different levels of deductible costs), or with effects of unobserved variables (such as genetic differences). Statistical techniques were also used to ascertain which results applied across the whole enrolled population and which were confined to certain subgroups (such as individuals with high blood pressure) and to translate utilization rates across different programs and types of patients into comparable overall dollar costs and health outcomes for alternative financing options.

A classical experiment, with systematic but randomly assigned variation of the variables of interest (or some reasonable approach to this), is usually considered the most rigorous basis from which to draw such inferences. But random samples or randomized experimental manipulations are not always feasible or ethically acceptable. Then, causal inferences must be drawn from observational studies, which, however well designed, are less able to ensure that the observed (or inferred) relationships among variables provide clear evidence on the underlying mechanisms of cause and effect.

Certain recurrent challenges have been identified in studying causal inference. One challenge arises from the selection of background variables to be measured, such as the sex, nativity, or parental religion of individuals in a comparative study of how education affects occupational success. The adequacy of classical methods of matching groups in background variables and adjusting for covariates needs further investigation. Statistical adjustment of biases linked to measured background variables is possible, but it can become complicated. Current work in adjustment for selectivity bias is aimed at weakening implausible assumptions, such as normality, when carrying out these adjustments. Even after adjustment has been made for the measured background variables, other, unmeasured variables are almost always still affecting the results (such as family transfers of wealth or reading habits). Analyses of how the conclusions might change if such unmeasured variables could be taken into account is essential in attempting to make causal inferences from an observational study, and systematic work on useful statistical models for such sensitivity analyses is just beginning.

The third important issue arises from the necessity for distinguishing among competing hypotheses when the explanatory variables are measured with different degrees of precision. Both the estimated size and significance of an effect are diminished when it has large measurement error, and the coefficients of other correlated variables are affected even when the other variables are measured perfectly. Similar results arise from conceptual errors, when one measures only proxies for a theoretical construct (such as years of education to represent amount of learning). In some cases, there are procedures for simultaneously or iteratively estimating both the precision of complex measures and their effect on a particular criterion.

Although complex models are often necessary to infer causes, once their output is available, it should be translated into understandable displays for evaluation. Results that depend on the accuracy of a multivariate model and the associated software need to be subjected to appropriate checks, including the evaluation of graphical displays, group comparisons, and other analyses.

New Statistical Techniques

Internal Resampling

One of the great contributions of twentieth-century statistics was to demonstrate how a properly drawn sample of sufficient size, even if it is only a tiny fraction of the population of interest, can yield very good estimates of most population characteristics. When enough is known at the outset about the characteristic in question—for example, that its distribution is roughly normal—inference from the sample data to the population as a whole is straightforward, and one can easily compute measures of the certainty of inference, a common example being the 95 percent confidence interval around an estimate. But population shapes are sometimes unknown or uncertain, and so inference procedures cannot be so simple. Furthermore, more often than not, it is difficult to assess even the degree of uncertainty associated with complex data and with the statistics needed to unravel complex social and behavioral phenomena.

Internal resampling methods attempt to assess this uncertainty by generating a number of simulated data sets similar to the one actually observed. The definition of similar is crucial, and many methods that exploit different types of similarity have been devised. These methods provide researchers the freedom to choose scientifically appropriate procedures and to replace procedures that are valid under assumed distributional shapes with ones that are not so restricted. Flexible and imaginative computer simulation is the key to these methods. For a simple random sample, the "bootstrap" method repeatedly resamples the obtained data (with replacement) to generate a distribution of possible data sets. The distribution of any estimator can thereby be simulated and measures of the certainty of inference be derived. The "jackknife" method repeatedly omits a fraction of the data and in this way generates a distribution of possible data sets that can also be used to estimate variability. These methods can also be used to remove or reduce bias. For example, the ratio-estimator, a statistic that is commonly used in analyzing sample surveys and censuses, is known to be biased, and the jackknife method can usually remedy this defect. The methods have been extended to other situations and types of analysis, such as multiple regression.

There are indications that under relatively general conditions, these methods, and others related to them, allow more accurate estimates of the uncertainty of inferences than do the traditional ones that are based on assumed (usually, normal) distributions when that distributional assumption is unwarranted. For complex samples, such internal resampling or subsampling facilitates estimating the sampling variances of complex statistics.

An older and simpler, but equally important, idea is to use one independent subsample in searching the data to develop a model and at least one separate subsample for estimating and testing a selected model. Otherwise, it is next to impossible to make allowances for the excessively close fitting of the model that occurs as a result of the creative search for the exact characteristics of the sample data—characteristics that are to some degree random and will not predict well to other samples.

Robust Techniques

Many technical assumptions underlie the analysis of data. Some, like the assumption that each item in a sample is drawn independently of other items, can be weakened when the data are sufficiently structured to admit simple alternative models, such as serial correlation. Usually, these models require that a few parameters be estimated. Assumptions about shapes of distributions, normality being the most common, have proved to be particularly important, and considerable progress has been made in dealing with the consequences of different assumptions.

More recently, robust techniques have been designed that permit sharp, valid discriminations among possible values of parameters of central tendency for a wide variety of alternative distributions by reducing the weight given to occasional extreme deviations. It turns out that by giving up, say, 10 percent of the discrimination that could be provided under the rather unrealistic assumption of normality, one can greatly improve performance in more realistic situations, especially when unusually large deviations are relatively common.

These valuable modifications of classical statistical techniques have been extended to multiple regression, in which procedures of iterative reweighting can now offer relatively good performance for a variety of underlying distributional shapes. They should be extended to more general schemes of analysis.

In some contexts—notably the most classical uses of analysis of variance—the use of adequate robust techniques should help to bring conventional statistical practice closer to the best standards that experts can now achieve.

Many Interrelated Parameters

In trying to give a more accurate representation of the real world than is possible with simple models, researchers sometimes use models with many parameters, all of which must be estimated from the data. Classical principles of estimation, such as straightforward maximum-likelihood, do not yield reliable estimates unless either the number of observations is much larger than the number of parameters to be estimated or special designs are used in conjunction with strong assumptions. Bayesian methods do not draw a distinction between fixed and random parameters, and so may be especially appropriate for such problems.

A variety of statistical methods have recently been developed that can be interpreted as treating many of the parameters as or similar to random quantities, even if they are regarded as representing fixed quantities to be estimated. Theory and practice demonstrate that such methods can improve the simpler fixed-parameter methods from which they evolved, especially when the number of observations is not large relative to the number of parameters. Successful applications include college and graduate school admissions, where quality of previous school is treated as a random parameter when the data are insufficient to separately estimate it well. Efforts to create appropriate models using this general approach for small-area estimation and undercount adjustment in the census are important potential applications.

Missing Data

In data analysis, serious problems can arise when certain kinds of (quantitative or qualitative) information is partially or wholly missing. Various approaches to dealing with these problems have been or are being developed. One of the methods developed recently for dealing with certain aspects of missing data is called multiple imputation: each missing value in a data set is replaced by several values representing a range of possibilities, with statistical dependence among missing values reflected by linkage among their replacements. It is currently being used to handle a major problem of incompatibility between the 1980 and previous Bureau of Census public-use tapes with respect to occupation codes. The extension of these techniques to address such problems as nonresponse to income questions in the Current Population Survey has been examined in exploratory applications with great promise.

Computing

Computer Packages and Expert Systems

The development of high-speed computing and data handling has fundamentally changed statistical analysis. Methodologies for all kinds of situations are rapidly being developed and made available for use in computer packages that may be incorporated into interactive expert systems. This computing capability offers the hope that much data analyses will be more carefully and more effectively done than previously and that better strategies for data analysis will move from the practice of expert statisticians, some of whom may not have tried to articulate their own strategies, to both wide discussion and general use.

But powerful tools can be hazardous, as witnessed by occasional dire misuses of existing statistical packages. Until recently the only strategies available were to train more expert methodologists or to train substantive scientists in more methodology, but without the updating of their training it tends to become outmoded. Now there is the opportunity to capture in expert systems the current best methodological advice and practice. If that opportunity is exploited, standard methodological training of social scientists will shift to emphasizing strategies in using good expert systems—including understanding the nature and importance of the comments it provides—rather than in how to patch together something on one's own. With expert systems, almost all behavioral and social scientists should become able to conduct any of the more common styles of data analysis more effectively and with more confidence than all but the most expert do today. However, the difficulties in developing expert systems that work as hoped for should not be underestimated. Human experts cannot readily explicate all of the complex cognitive network that constitutes an important part of their knowledge. As a result, the first attempts at expert systems were not especially successful (as discussed in Chapter 1). Additional work is expected to overcome these limitations, but it is not clear how long it will take.

Exploratory Analysis and Graphic Presentation

The formal focus of much statistics research in the middle half of the twentieth century was on procedures to confirm or reject precise, a priori hypotheses developed in advance of collecting data—that is, procedures to determine statistical significance. There was relatively little systematic work on realistically rich strategies for the applied researcher to use when attacking real-world problems with their multiplicity of objectives and sources of evidence. More recently, a species of quantitative detective work, called exploratory data analysis, has received increasing attention. In this approach, the researcher seeks out possible quantitative relations that may be present in the data. The techniques are flexible and include an important component of graphic representations. While current techniques have evolved for single responses in situations of modest complexity, extensions to multiple responses and to single responses in more complex situations are now possible.

Graphic and tabular presentation is a research domain in active renaissance, stemming in part from suggestions for new kinds of graphics made possible by computer capabilities, for example, hanging histograms and easily assimilated representations of numerical vectors. Research on data presentation has been carried out by statisticians, psychologists, cartographers, and other specialists, and attempts are now being made to incorporate findings and concepts from linguistics, industrial and publishing design, aesthetics, and classification studies in library science. Another influence has been the rapidly increasing availability of powerful computational hardware and software, now available even on desktop computers. These ideas and capabilities are leading to an increasing number of behavioral experiments with substantial statistical input. Nonetheless, criteria of good graphic and tabular practice are still too much matters of tradition and dogma, without adequate empirical evidence or theoretical coherence. To broaden the respective research outlooks and vigorously develop such evidence and coherence, extended collaborations between statistical and mathematical specialists and other scientists are needed, a major objective being to understand better the visual and cognitive processes (see Chapter 1) relevant to effective use of graphic or tabular approaches.

Combining Evidence

Combining evidence from separate sources is a recurrent scientific task, and formal statistical methods for doing so go back 30 years or more. These methods include the theory and practice of combining tests of individual hypotheses, sequential design and analysis of experiments, comparisons of laboratories, and Bayesian and likelihood paradigms.

There is now growing interest in more ambitious analytical syntheses, which are often called meta-analyses. One stimulus has been the appearance of syntheses explicitly combining all existing investigations in particular fields, such as prison parole policy, classroom size in primary schools, cooperative studies of therapeutic treatments for coronary heart disease, early childhood education interventions, and weather modification experiments. In such fields, a serious approach to even the simplest question—how to put together separate estimates of effect size from separate investigations—leads quickly to difficult and interesting issues. One issue involves the lack of independence among the available studies, due, for example, to the effect of influential teachers on the research projects of their students. Another issue is selection bias, because only some of the studies carried out, usually those with "significant" findings, are available and because the literature search may not find out all relevant studies that are available. In addition, experts agree, although informally, that the quality of studies from different laboratories and facilities differ appreciably and that such information probably should be taken into account. Inevitably, the studies to be included used different designs and concepts and controlled or measured different variables, making it difficult to know how to combine them.

Rich, informal syntheses, allowing for individual appraisal, may be better than catch-all formal modeling, but the literature on formal meta-analytic models is growing and may be an important area of discovery in the next decade, relevant both to statistical analysis per se and to improved syntheses in the behavioral and social and other sciences.

OPPORTUNITIES AND NEEDS

This chapter has cited a number of methodological topics associated with behavioral and social sciences research that appear to be particularly active and promising at the present time. As throughout the report, they constitute illustrative examples of what the committee believes to be important areas of research in the coming decade. In this section we describe recommendations for an additional $16 million annually to facilitate both the development of methodologically oriented research and, equally important, its communication throughout the research community.

Methodological studies, including early computer implementations, have for the most part been carried out by individual investigators with small teams of colleagues or students. Occasionally, such research has been associated with quite large substantive projects, and some of the current developments of computer packages, graphics, and expert systems clearly require large, organized efforts, which often lie at the boundary between grant-supported work and commercial development. As such research is often a key to understanding complex bodies of behavioral and social sciences data, it is vital to the health of these sciences that research support continue on methods relevant to problems of modeling, statistical analysis, representation, and related aspects of behavioral and social sciences data. Researchers and funding agencies should also be especially sympathetic to the inclusion of such basic methodological work in large experimental and longitudinal studies. Additional funding for work in this area, both in terms of individual research grants on methodological issues and in terms of augmentation of large projects to include additional methodological aspects, should be provided largely in the form of investigator-initiated project grants.

Ethnographic and comparative studies also typically rely on project grants to individuals and small groups of investigators. While this type of support should continue, provision should also be made to facilitate the execution of studies using these methods by research teams and to provide appropriate methodological training through the mechanisms outlined below.

Overall, we recommend an increase of $4 million in the level of investigator-initiated grant support for methodological work. An additional $1 million should be devoted to a program of centers for methodological research.

Many of the new methods and models described in the chapter, if and when adopted to any large extent, will demand substantially greater amounts of research devoted to appropriate analysis and computer implementation. New user interfaces and numerical algorithms will need to be designed and new computer programs written. And even when generally available methods (such as maximum-likelihood) are applicable, model application still requires skillful development in particular contexts. Many of the familiar general methods that are applied in the statistical analysis of data are known to provide good approximations when sample sizes are sufficiently large, but their accuracy varies with the specific model and data used. To estimate the accuracy requires extensive numerical exploration. Investigating the sensitivity of results to the assumptions of the models is important and requires still more creative, thoughtful research. It takes substantial efforts of these kinds to bring any new model on line, and the need becomes increasingly important and difficult as statistical models move toward greater realism, usefulness, complexity, and availability in computer form. More complexity in turn will increase the demand for computational power. Although most of this demand can be satisfied by increasingly powerful desktop computers, some access to mainframe and even supercomputers will be needed in selected cases. We recommend an additional $4 million annually to cover the growth in computational demands for model development and testing.

Interaction and cooperation between the developers and the users of statistical and mathematical methods need continual stimulation—both ways. Efforts should be made to teach new methods to a wider variety of potential users than is now the case. Several ways appear effective for methodologists to communicate to empirical scientists: running summer training programs for graduate students, faculty, and other researchers; encouraging graduate students, perhaps through degree requirements, to make greater use of the statistical, mathematical, and methodological resources at their own or affiliated universities; associating statistical and mathematical research specialists with large-scale data collection projects; and developing statistical packages that incorporate expert systems in applying the methods.

Methodologists, in turn, need to become more familiar with the problems actually faced by empirical scientists in the laboratory and especially in the field. Several ways appear useful for communication in this direction: encouraging graduate students in methodological specialties, perhaps through degree requirements, to work directly on empirical research; creating postdoctoral fellowships aimed at integrating such specialists into ongoing data collection projects; and providing for large data collection projects to engage relevant methodological specialists. In addition, research on and development of statistical packages and expert systems should be encouraged to involve the multidisciplinary collaboration of experts with experience in statistical, computer, and cognitive sciences.

A final point has to do with the promise held out by bringing different research methods to bear on the same problems. As our discussions of research methods in this and other chapters have emphasized, different methods have different powers and limitations, and each is designed especially to elucidate one or more particular facets of a subject. An important type of interdisciplinary work is the collaboration of specialists in different research methodologies on a substantive issue, examples of which have been noted throughout this report. If more such research were conducted cooperatively, the power of each method pursued separately would be increased. To encourage such multidisciplinary work, we recommend increased support for fellowships, research workshops, and training institutes.

Funding for fellowships, both pre- and postdoctoral, should be aimed at giving methodologists experience with substantive problems and at upgrading the methodological capabilities of substantive scientists. Such targeted fellowship support should be increased by $4 million annually, of which $3 million should be for predoctoral fellowships emphasizing the enrichment of methodological concentrations. The new support needed for research workshops is estimated to be $1 million annually. And new support needed for various kinds of advanced training institutes aimed at rapidly diffusing new methodological findings among substantive scientists is estimated to be $2 million annually.

6

The Research Support System

6

The Research Support System

Previous chapters of this report focused on the substance of the behavioral and social sciences: What are the major questions and ideas that drive research and give shape to the fields? What advances are occurring in the methods by which new knowledge is discovered and validated? But some conditions affect opportunities for substantive progress across all of the topics, and so in this chapter we step back and consider the institutional context in which the work is conducted—the research support system.

The elements of the system can be characterized by the resources on which research depends:

- **Human resources.** People become researchers through programs of teaching, training, and professional certification in academic departments, affiliated or independent institutes, and professional associations.

- **Technological resources.** These resources depend on programs of technical support, maintenance, and procurement to ensure that researchers have high-quality, fully operational laboratories, field instruments, computers, specialized facilities, communications systems to facilitate knowledge and collaboration, and supplies.

- **Data resources.** Of particular concern here are large-scale data sets, including those that are part of the federal statistical system; specially collected research data; and research-relevant record systems kept by other private or public organizations principally for administrative, management, or intelligence purposes.

■ *Funding resources.* The availability of public and private funds is determined by the financial commitments of funding organizations and the procedures used by them to solicit, review, monitor, and coordinate expenditures through intramural and extramural budget allocations, grants, and contracts.

These four resource categories provide a useful organizing framework, although the institutions that support research cannot be divided neatly by category. For example, colleges and universities consider teaching a primary responsibility, but they also take responsibility for maintaining technological resources needed for research. Many academic institutions also maintain data archives, central computer facilities, or specialized research organizations, and many allocate funds to support faculty and student research. Similarly, the National Science Foundation is known principally as a research funding agency, but its Division of Science Resource Studies is a major collector, analyst, and source of data on scientific and engineering personnel, facilities, and expenditures, covering federal and state agencies, other countries, and the private sector. And the National Institute of Mental Health not only funds a substantial amount of extramural research, but also houses some of the world's top laboratory facilities and scientific talent.

This overlay of roles applies even to many rather specialized behavioral and social sciences institutions. For example, the Inter-university Consortium for Political and Social Research, headquartered at the Institute for Social Research at the University of Michigan but with nearly 300 member institutions, is a central archive for machine-readable research data, a center for training in quantitative techniques of analysis, and a source of technical computing assistance and software development. In short, institutions often play more than one role in supporting research.

There are a number of opportunities for improving the research support system of the behavioral and social sciences. Each section in this chapter begins with an overview of the resource situation and then provides recommendations for change. The recommendations are not addressed to any one institution; they call for the cooperative and imaginative efforts of several. Even when a single focus seems apparent—such as changes in the activities of government funding agencies—change is not simply a matter of deciding internally on new policy initiatives. Funding agencies rely very heavily on the research community to generate proposals for research and to provide critical technical evaluations, priority ratings, and program guidance. Those evaluations play a large role in determining what research is supported, at what level of effort and for what length of time, and how funds are allocated among individual projects, multipurpose equipment, collective data resources, and investments in human capital. Moreover, government funding agencies need to justify their programs and budgets to policy makers in Congress and the executive branch, who are also responsive to the research community. Consequently, the analysis and recommendations in this chapter are as important to researchers as to administrators, and addressed as much to universities and other research institutions as to foundations and other sponsors.

HUMAN RESOURCES

Measured by the number of PhDs (and equivalent research doctorates), the behavioral and social sciences are large. About 120,000 people in the United States hold doctoral degrees, and about 6,000 new PhDs are granted annually in these fields. These fields generally include anthropology, economics, geography, linguistics, political science, psychology, sociology, statistics, and closely related fields such as criminology and international relations. The numbers do not include the fields of history—about 500 doctorates in 1986—or education, in which the subfields of educational psychology, educational statistics, and educational testing granted about 500 EdDs in 1986.

This numerical strength does not translate into proportionate strength in research. Although behavioral and social scientists constitute 30 percent of all science and engineering doctorates, they constitute only 13 percent of those whose primary activity is research and development. On university and college campuses, behavioral and social scientists comprise 22 percent of all the full-time equivalent scientists and engineers, but only 8 percent of the 65,000 full-time equivalent positions devoted to research and development. This last percentage is lower than it was a decade ago, in approximate proportion to a reduction in federal research support. But even at its highest point, campus strength in behavioral and social sciences research was well short of what one might expect on the basis of the numbers of trained personnel available. An explanation and a prescription for this persistent difference must begin by taking into account the balance between general education and research training in the behavioral and social sciences in universities and colleges.

Colleges and Graduate Schools

Undergraduate Education

The current structure of curricula, beginning in secondary school, leads to teaching demands that strongly affect the research productivity of talented young academic scientists in the behavioral and social sciences. We recommend that senior decision makers in universities and elsewhere support long-term programmatic and procedural changes to upgrade the level of behavioral and social sciences curricula in secondary schools and in colleges.

A very large number of college students take courses and major in the behavioral and social sciences; about 110,000 bachelor's degrees are granted annually to behavioral and social sciences majors. Although substantially less than the peak of 145,000 in 1974, the number is still approximately equal to the number of bachelor's degrees granted annually to physical, mathematical, and life sciences majors combined. The relative popularity of behavioral and social sciences undergraduate courses results in heavy teaching demands in these fields: more than one-fourth of doctoral psychologists and one-half of all other social scientists report that teaching, largely undergraduate instruction, is their principal activity.

Most students in introductory, lower-division courses in the behavioral and social sciences have had no opportunity to take precollege courses in these disciplines. They have gained a certain amount of knowledge in less systematized or differently organized courses in grade school social studies and in high school civics, economics, world geography, U.S. and world history, health and safety, and so on. But such courses seldom convey a sense of the scientific nature of inquiry, teach the theoretical foundations of knowledge, or provide basic lessons in research methods.

This situation contrasts sharply with the experience of introductory students in such scientific fields as biology or physics: first, as a college requirement, many of them have taken specific disciplinary courses at the high school level; second, those courses include lessons in basic theory and hands-on experience with laboratory research methods. One consequence of this sharp difference is that high school students with talents and interests suitable for formal scientific or technical training are less likely to be aware of opportunities to pursue such intellectually challenging careers in the behavioral and social sciences. And in fact, students who major in the behavioral and social sciences are much less likely than students majoring in other scientific fields to continue on to graduate school. In addition, students in lower-division college courses in the behavioral and social sciences have usually done little or no work in mathematical statistics, decision theory, networks, or other mathematical subjects particularly relevant to the behavioral and social sciences.

Past efforts in science education, sponsored by the National Science Foundation among others, have had much to do with upgrading high school curricula in mathematics and the physical and life sciences. Virtually no such efforts have been undertaken at the federal level to upgrade precollege instruction in the behavioral and social sciences, though some fledgling attempts have been made at the state level.

Lower-division instruction in the behavioral and social sciences is, therefore, necessarily different in character from that in the physical and life sciences. In the behavioral and social sciences, instructors must first focus on defining the basic subject matter, recasting students' earlier, ad hoc knowledge into systems of disciplinary thought, and giving them rudimentary training in relevant research methods. In the natural sciences, in contrast, introductory courses can focus on deepening students' understanding of previously introduced theoretical ideas and expanding their methodological skills.

Overall, greater demand is placed on behavioral and social sciences professors in universities and colleges to prepare students for nonresearch careers. An appearance of similarity in teaching loads (as measured, say, by number of students) between faculties often masks an important underlying difference: an imbalance between the number of students in lower- and upper-division courses. At many universities, the physical and life sciences faculty teach relatively fewer courses, and most of the ones taught are either large introductory surveys or small laboratories or seminars. In the behavioral and social sciences, the lower-division survey courses are smaller and upper-division seminars and practicum courses are larger. The substantial difference between the numbers of students in lower- and upper-division courses allows intense professorial attention to students at the upper level. Instruction for physical and life science students at this level is often aimed at preparing them for scientific or technical careers. This contrast at upper undergraduate levels between the faculty-intensity and research orientation of natural science education and the broader intellectual net that is cast to serve the needs of less research-oriented undergraduate majors in the behavioral and social sciences may make the latter less attractive to many intellectually talented undergraduates.

The features of the U.S. educational system that create these differences are not likely to change rapidly. Hence, the teaching burden on young behavioral and social sciences professors can be expected to continue. Several of our recommendations are designed primarily to insulate more of the best young academic scientists from the systematic pressures that inhibit high-quality, productive research. But in the long term the upgrading of curriculum at the secondary school and undergraduate levels would pay large dividends in the quality and quantity of research.

Graduate Education

The research opportunities identified in this report are threatened by the significant declines since the mid-1970s in federal and other support for graduate work, especially at research-oriented universities. We recommend that a tightly focused and clearly articulated effort be undertaken immediately to attract greater proportions of research-oriented students into graduate programs in the behavioral and social sciences and to intensify and upgrade the average level of research training in graduate schools. We recommend $10 million in new funding for predoctoral fellowships and training grants. This amount represents close to one-fourth of all the new human resource investments recommended in this report.

The most serious concern of graduate departments in the behavioral and social sciences, especially those in which training for research takes high prior-

TABLE 6-1   Support for Graduate Students in the Behavioral and Social Sciences at Doctorate-Granting Universities, 1975 and 1985

	Number of Students	
Kind of Support	1975	1985
Research assistantships		
All sources	7,173	8,102
Federal sources	2,441	1,962
Fellowships and traineeships		
All sources	13,559	10,124
Federal sources	4,595	2,313

SOURCE: Data from National Science Foundation, *Science Indicators: The 1985 Report*, NSB85-1 and *Academic Science/Engineering: Graduate Enrollment and Support*, Fall 1985, SRS87-D5.

ity, is the increasing difficulty of recruiting and retaining talented graduate students with a commitment to research. Overall, graduate enrollments and the number of doctorates completed in the behavioral and social sciences have been fairly constant over the past decade, but the number of doctorates completed in departments ranking in the top quartile of scientific quality (as indexed in a 1982 study by the Conference Board of Associated Research Councils) declined by 17 percent between 1973 and 1983; this decline is more than twice the decline in the number of doctorates granted by the top quartile of all other science and engineering departments.

While there are many reasons for the decrease in completed doctorates at top research departments, one major contribution is undoubtedly the retrenchment of federal support (see Table 6-1). Between 1975 and 1985, federal support for graduate research assistants, fellows, and trainees decreased significantly. The decrease in support for assistants was more than compensated for by increases from academic or other sources (though the average value of stipends may have fallen), but this was not the case for fellows and trainees. Although federal support for them decreased substantially across all the fields of science, it was steepest, 2,250 or 49 percent of all positions, in the behavioral and social sciences—the fields that were much more dependent than other sciences on that support. An additional 1,000 nonfederally funded graduate positions disappeared during the same years. The decline was especially evident in the most prominent research departments. For example, between 1975 and 1982, the total number of full-time graduate students receiving any kind of federal support in the top quartile of behavioral and social sciences departments declined by 53 percent. In other scientific fields, the comparable number *increased* by 15 percent.

To support their graduate training, behavioral and social sciences graduate students now rely heavily on their own earnings, spouses' earnings, and loans (see Table 6-2). Moreover, the assistance that is available is skewed toward providing teaching assistants for lower-division instruction. From the perspective of research training, the most important point is that since there is limited financial assistance for actual research, the proportion of time devoted to developing research skills and interests is similarly limited. As a result, completion of graduate training in the behavioral and social sciences is relatively slow, about 2 years longer than in natural sciences. There are also more part-time students, attrition is higher, and new PhDs are generally older.

If present patterns of support continue, it will be difficult to sustain the base of talented and committed young scientists needed to keep the national research enterprise competitive and enable it to exploit the scientific opportunities discussed in this report. It is especially important that high-quality graduate fellowships, traineeships, and assistantships once again become readily available in conjunction with faculty research and advanced training. Only by paying serious attention to the financial requirements of graduate training will the behavioral and social sciences stop losing potential research talent to clinical, business, legal, and other kinds of career training.

In the short run, the most significant improvements will be achieved by increased support for predoctoral research fellowships and training centers. Support of graduate training in the form of national competitive fellowships has two functions: it provides incentives for individuals to enter the fields, and it provides information to undergraduate students about the kinds of training needed to enter those fields. Support of fellowship programs for graduate study in the behavioral and social sciences should be increased with the specific intent of encouraging more research-oriented students and those with rigorous undergraduate backgrounds—including those with majors in the natural sciences or formal methodological disciplines, such as mathematics and logic, computer science, and statistics—to undertake such study.

The quality of graduate training can also be influenced significantly through the support of training programs that require students to go beyond the stand-

TABLE 6-2 Sources of Support for Graduate Studies Reported by Doctoral Recipients in 1986, by Field (percentage)

Source of Support	Life Sciences	Physical Sciences	Engineering	Behavioral and Social Sciences
Own earnings	46	34	38	68
Spousal earnings	31	21	17	33
Loans	33	20	14	59
Teaching assistantships	40	70	43	57
Research assistantships	54	74	72	38

NOTE: Columns add to more than 100 percent due to multiple sources of support.

SOURCE: Data from *Summary Report 1986: Doctorate Recipients From United States Universities.* Office of Scientific and Engineering Personnel, National Research Council. National Academy Press, 1987.

ard requirements of their disciplines, either across the fields of the behavioral and social sciences or in methodological disciplines. As noted elsewhere, university departments are organized primarily along disciplinary lines, and almost all graduate training takes place within disciplines. One of the costs of such departmental structures is that interdisciplinary research and training is overtly or covertly discouraged. The burden of proof as to the value of pursuing such interests falls on those who want to cross departmental boundaries. We believe that improvements in training and advances in research can be facilitated by crossing those boundaries. Universities, colleges, and support agencies should seek ways to reinforce activities that point toward promising areas of interdisciplinary and collaborative research. Graduate students should be encouraged to span departmental programs and participate in multidisciplinary research on campuses as freely as they undertake disciplinary activities.

Postdoctoral Training and Collaboration

Despite graduate-level financial support problems, a substantial number of students do eventually complete PhDs in the behavioral and social sciences. But it is neither easy nor typical for these new PhDs to enter research careers. In comparison with other fields of science, there are few entry-level behavioral and social sciences research positions available. In 1986, for the 5,700 new life science PhDs, there were about 3,750 postdoctoral research positions (fellowships, associateships) and entry-level jobs primarily devoted to research and development, about 66 prime research openings per 100 doctorates For the 8,200 new physical and engineering sciences PhDs in 1986, there were about 4,850 postdoctoral research positions and research and development jobs, about 59 openings per 100 doctorates. For the 5,850 new behavioral and social sciences PhDs in 1986, there were only about 1,600 such fellowships and jobs, about 27 openings per 100 doctorates. Adding jobs in which research and development is a secondary activity, the ratios of openings to new PhDs are 73, 68, and 43 per 100 for life science, physical and engineering sciences, and behavioral and social sciences, respectively.

These differences parallel substantial differences in academic culture among new faculty at universities and colleges. As explained above, the teaching loads of assistant professors in the behavioral and social sciences are often heavier than in other scientific fields. Universities generally do not provide research funds or facilities for behavioral and social sciences faculty as routinely as they do for other new sciences faculty. Except in one or two fields, generally lower salaries put financial pressure on new faculty to supplement their standard 9-month teaching base pay. As a result, unless they can garner immediate grant support, "spare" time must be devoted to summer teaching or similar employment instead of to research.

Thus, the most crippling constraint on research for most new (and mid-career) scientists, for professors as well as for those in other jobs, is a lack of time, and time is critical for research. Time must be found for writing proposals, organizing laboratory or data resources, thinking through experiments or analytical strategies, reading the relevant literatures, recruiting and training assistants, consulting with colleagues, and eventually, writing up results, interpretations, and theories. There is no single model of support to provide more time for research: released time during one or more academic years, summer support, and full-time leaves of absence for periods of weeks, months, or several years may all be appropriate in different instances. Other sources of support for research are advanced workshops away from a person's home institution, regional and national research centers, and traditional postdoctoral fellowships.

The central point is that new PhDs as well as those early in their careers need to have access to a range of postdoctoral research opportunities that are not necessarily tied to specific grants for fully described projects. These opportunities must be compatible with the career needs and research interest of young scientists. An increase in financial resources is needed, but an expansion of imaginative design of research possibilities is nearly as important as increasing the level of funds.

Postdoctoral Fellowships and Traineeships

We recommend an increase in the number of postdoctoral fellowships in the behavioral and social sciences, to bring the availability of postdoctoral opportunities more closely in line with research needs and opportunities. Special attention should be given to those research areas for which advanced training in more than one discipline is essential. We recommend an aggregate annual increase of $18 million for postdoctoral support, which should be divided so that there is a balance between new training grants to institutions, individual fellowships at the junior level, and more advanced fellowships. In addition, the timing of award decisions should be changed: awards to potential fellows, especially at the entry level, should be made early in the academic year to maximize their attractiveness.

One of the most successful devices to increase training, accumulate experience, and strengthen a scientist's research productivity is the position of the postdoctoral fellow or trainee. (As a rule, postdoctoral trainees are selected by the administrators of multiyear institutional training grants, and fellows are selected by direct application to the funding agency.) Fellowships are thus usually portable—attached to an individual, not to a particular training program. Such a position may be for as long as three years, and in some cases leads to a second advanced degree.

Present arrangements for strong postdoctoral programs in the behavioral and social sciences are unsatisfactory in at least two ways. The most obvious is that funds for postdoctoral training in the behavioral and social sciences are too small. Second, the timing of the cycle for application, evaluation, and award of such fellowships is not always conducive to attracting the best candidates. Most colleges and universities extend offers for new instructors and assistant professors between January and March; postdoctoral awards are typically offered later in the year. Many promising candidates therefore accept primary teaching positions that they might not have taken if an attractive postdoctoral fellowship had been available. As a result, the candidate pool for fellowships is reduced in numbers and quality.

Advanced Training Institutes

We recommend new support for short-duration advanced training institutes, generally of about 3 to 6 weeks in the summer months. Funding agencies should renew their receptivity to such proposals, which should be reviewed competitively with each other and separately from individual-investigator proposals. We recommend initiation of approximately 35 programs per summer of 25 participants each, at an estimated aggregate annual cost of $7 million.

During the 1950s and 1960s, numerous advanced training institutes (3 to 6 weeks in duration) were initiated; many centered on the teaching and refinement of mathematical and statistical techniques useful in the behavioral and social sciences. Those who participated—both as teachers and students—found these programs to be unusually successful. New methods were brought much more rapidly into use, new lines of research developed, important results were published, and people who were otherwise isolated from parts of their own or closely related research fields were brought effectively and efficiently up-to-date because of the close interaction that takes place in such a setting.

Most of these activities were discontinued, not because the need for them diminished but because of changes in the mode of funding. During the era in which they flourished, special research funds were available to institutions for such institutes. Evaluation of specific course proposals was conducted by the grantee institution, which had overall responsibility for hosting and providing organizational support for the activities, which were usually held in the summer. Thus, with a modest effort, it was possible for a group of researchers to develop a proposal for an institute and have it evaluated promptly and in competition only with similar activities. When the special funds were terminated, these proposals had to compete directly with all other research proposals. As a result, researchers had to write far more elaborate justifications to conform to standard investigator-type proposal formats and to wait longer for decisions through the centralized review system. They were often disappointed: when proposal evaluators confront direct choices between individual-investigator projects and those that involve collective and necessarily more complex activity, the former are usually favored. In this instance, a shift in the structure of funding led to the virtual abandonment of a very useful set of activities in nearly all behavioral and social sciences research areas.

The value of advanced training institutes justifies their resumption. There are many examples of areas in which training institutes would be very valuable: innovations in statistical methodology, some of which are powerful but not widely used; theories and applications of measurement and scaling; theoretical linkages across disciplinary boundaries, such as the interface between the neurosciences and cognitive sciences, decision making under risk and uncertainty, and the history and sociology of modern science (including the behavioral and social sciences); demographic and sociological aspects of the life cycle; new biological assay techniques; and advances in computer simulation models and their applications.

Collaboration and Communication

We recommend a special initiative to facilitate greater communication and collaboration among dispersed scholars working on substantively related research. We recommend that 45 new workshop programs be initiated, at an average cost of about $200,000 for a total of about $9 million per year in new funds. Moreover, research proposals that include requests to fund collaborative efforts, if recommended for award through competitive evaluation, should be provided sufficient funds to ensure that their collaborative features are retained.

Behavioral and social sciences research is characterized by decentralization both within universities and geographically across the country. Most academic departments, seeking intellectual breadth and balance to ensure adequacy in the undergraduate and graduate curriculum, have only one or a small number of faculty members with expertise in any given specialized line of inquiry. To achieve a localized intellectual critical mass is therefore difficult. One way to do so is to create research centers that can facilitate communication and collaboration among scientists, which is discussed later in this chapter. Yet less expensive and more flexible means than organized research centers are often sufficient to strengthen communication and collaboration among investigators who need to regularly exchange research ideas and methodological knowledge with colleagues located at other institutions.

The need for exchange calls for a well-developed, effective system of national and cross-national research networks to bring people together regularly or on occasion, for periods of 1 to 6 weeks. A few such arrangements do exist in some areas of the behavioral and social sciences, but these hardly constitute a system. Travel grant funds that in the past facilitated such collaboration in one or another field for some period have been drastically reduced in recent years. When the costs for a proposal are scrutinized, travel funds are often the first point of attack in trying to reduce costs. This practice of treating travel funds as expendable is counterproductive to advancing needed collaborative research. Other mechanisms for interdisciplinary communication, advanced training, and collaboration, such as specialized institutes, conferences, and individual travel grants, have also come under severe budgetary pressure. Continued progress at many research frontiers stands in need of new and renewed commitment to the promotion of intradisciplinary and interdisciplinary communication, training, and collaboration beyond the confines of home departments.

The need for collaborative working arrangements has several aspects: programs of several investigators working together on a joint project; workers on mutually relevant but not directly related projects keeping closely apprised of and contributing collegially to each other's work; research scientists in different countries coordinating, comparing, and advancing their theoretical and empirical work. The mechanism of intensive research workshops, in which groups of active investigators gather, usually for 1 to 2 weeks during the academic year and 2 to 4 weeks in the summer, has been very fruitful in the few behavioral and social sciences areas in which it has been used. Such programs should be intended to run for a minimum of 5 years and to be open to competitive renewal.

TECHNOLOGICAL RESOURCES

The behavioral and social sciences have long made use of certain kinds of scientific and technical equipment, including computers, behavioral and neuropsychological sensors, and sound and picture recording and display devices. However, a substantial part of this technological force has been weakened in the past decade. The inability to meet material and technological needs, particularly to replace old or obsolete equipment, is now a major bottleneck on the rate of scientific progress in some research areas. Particularly notable in all areas is the scarcity of powerful workstations and superminicomputers, which greatly exceed the capacities of personal computers but can readily be used as desktop instruments for departments or medium-sized research groups.

Recent figures regarding technological renewal in the behavioral and social sciences provide little encouragement (see Table 6-3). During fiscal 1981–1985, federal expenditures for research equipment in psychology increased 39 percent and nonfederal expenditures in psychology increased 47 percent. But these increases were below the overall 58 percent increase for all the sciences; moreover, expenditures for social science research equipment showed no net increase. Altogether, only 2.7 percent of the total $655 million spent for research equipment on campuses in fiscal 1985 was for psychology and the social sciences, down from 3.7 percent in 1981.

There is a persisting view that behavioral and social sciences research can operate as a virtually equipment-free enterprise, a view that is completely out of date for research in many areas. Some examples of equipment needs discussed in prior chapters include special video laboratories for the experimental study of human communication, interaction, and economic exchange; neu-

TABLE 6-3 Expenditures for Research Equipment at Universities and Colleges ($ million)

Science and Year	Total	Federal	Nonfederal
Psychology			
1981	6.0	4.4	1.6
1982	5.7	4.2	1.5
1983	6.6	4.8	1.9
1984	7.0	5.0	2.0
1985	8.5	6.2	2.3
Social sciences			
1981	9.2	4.0	5.2
1982	7.1	2.9	4.2
1983	9.0	2.9	6.0
1984	12.7	3.5	9.2
1985	9.3	4.2	5.1
All sciences			
1981	414.0	265.0	148.7
1984	518.1	335.5	182.6
1985	654.8	427.2	227.7

SOURCE: Data from National Science Foundation, *Science Indicators: The 1985 Report,* NSB 85-1, and *Academic Science/Engineering: R&D Funds, Fiscal Year 1985.*

roimaging technology to identify structural features and metabolically active regions in the living brain; and equipment to analyze satellite sensing of geographical features. There has been a tendency to neglect these and other technological needs of the behavioral and social sciences, even to the point of sometimes not including them from in national surveys of equipment resource needs in scientific facilities. Detailed analytical assessments of equipment needs in the behavioral and social sciences and year-to-year planning to meet the needs discerned are urgently needed.

Computers

We recommend $22 million annually in new funds to purchase and support improved computational technology. This amount is almost one-half of the $51 million total recommended in this report for technological resources. Centralized assistance in preparing, documenting, and providing software suited to behavioral and social sciences problems is needed; more particularly, programmers working in or trained at centralized computer facilities must be familiar with such special software libraries, needs, and problems. Access to advanced computing facilities by behavioral and social scientists should be improved. Behavioral and social scientists need to be involved directly in the planning, assessment, and development of new supercomputer facilities.

The behavioral and social sciences have a long history of using electronic computational devices for experimental control, data collection, data analysis, and designing and testing models of social and individual behavior. Behavioral and social sciences researchers have also had leading roles in the design of user-friendly computer systems and special computer tools to aid human-computer interaction. The shift from mainframe computers to powerful desktop machines, including new generations of very powerful workstations, has opened entirely new possibilities for rapid, real-time data analysis, sophisticated model building, and networking. The possibilities opened by electronic communication networks have barely been tapped and will require much greater future investments in suitable technology. Although funding for the purchase of computing hardware or the development or purchase of software tailored for behavioral and social sciences is growing, the need far exceeds the available funds.

The currently available computer hardware and software are already inadequate to meet the analytic, management, and retrieval needs of some of the larger survey data collections. As these data collections continue to grow, the adequacy of current computer facilities will be still further strained. The development of supercomputers is a possible solution to meeting the requirements of users of the very largest data bases, but it is by no means certain. As development is currently planned, supercomputers will have the capacity to carry out very large numbers of calculations at very high speeds, which have great advantages in meeting the mathematical needs of complex iterative modeling, but they may not have capacities to manipulate large quantities of data efficiently (input-output capacities) and thus to handle analyses of large-scale data bases.

Behavioral and social sciences research is straining the limits of present computational resources in several other areas, such as: the theoretical simulation of complex systems of simultaneous processes, for example, computational models of the visual system, brain functioning at both the cognitive and neurophysiological levels, or human-computer interactions; and the empirical estimation, from even moderately large data sets, of the coefficients of complex systems of equations used in psychometric and test theory, scaling procedures, and models of national and international economies. Statistical covariance structure models, log-linear models of categorical data, nonmetric multidimensional scaling, and special clustering and classification models are suffering from computational overload with present facilities. Greater use of advanced workstations and supercomputers would also encourage better analyses of nonnormal multivariate distributions in nonexperimental data by such powerful techniques as structural equations, canonical correlation, principal components, and discriminant functions. Use of these techniques for nonnormal data requires computations several magnitudes more extensive than those needed for normal models.

Currently, the natural scale of many scientific problems must be reduced to fit available computing machinery, in one of two ways. The investigator can restrict the analysis to a smaller or less precisely specified aspect of the system under study, but thereby lose power to discriminate between alternative theories. Or the investigator can subdivide the problem into pieces that can be separately carried out on a conventional computer, but thereby lose the ability to optimize an overall solution or work with any rapidity. A better solution would be possible if the investigator could gain access to Class VI supercomputers, such as the Cray-1 and Cyber 205, or to use the capabilities of connectionist (non-von Neumann) architectures, such as the new large-scale parallel-processing computers. Parallel architectures appear to be valuable for many applications in cognitive science, including connectionist models, computational linguistics, and the study of problem solving and theorem proving, and in macroeconomics for the simulation and analysis of multicountry and global dynamic models.

An issue within universities is the allocation of mainframe computer costs. Decisions on how costs are allocated for central processor time, blocks of memory, and input/output channels can significantly affect computer use. For example, many university computer centers have high charges for memory storage and input/output channel use, which means that users whose work involves transformation or combination of large data bases pay far more to run their jobs than users with small data bases but using complex equation systems or iterative algorithms that require a great deal of straight computation. Yet there is little difference between the two jobs in terms of incremental cost to the facility. With grant sizes decreasing and computer costs increasing, researchers whose needs happen to be high on the wrong index of mainframe user-cost accounting may have inadequate resources to do high-quality work. The solution to this problem is to revise computer user-cost systems that create inadvertent cross-subsidies.

Visual perception research has an especially rich history of advances correlated with progress in computer graphic capabilities. Important research questions can best be addressed with experiments involving the rapid generation of medium- to high-resolution naturalistic images. Prerequisites to effective computer use in visual perception include wide diffusion of new software resources, appropriate local hardware (such as wide bandwidth communication devices, graphics workstations, and image processors), and collaboration between visual science and computer graphics specialists. In addition, any field in which new computer models or methods are being developed needs rapid, cost-efficient ways to transmit and test programs and procedures among different sites. Sharing of programs is crucial in order to permit scientists to experiment with each other's theoretical tools; appropriate networks are essential.

In summary, the importance of advanced scientific computing to progress on many behavioral and social science research problems needs to be recognized. Support must be available for use of Class VI machines and other architectures, symbolic and parallel-processing capabilities, high-capacity communication channels, scientific computing networks, and local needs including new workstations and software development.

Neuroimaging Devices

We recommend an initiative to ensure access to neuroimaging devices for nonclinical researchers studying language, memory, or motivational phenomena. We recommend that $4 million be provided annually for this purpose.

Some of the neurolinguists, neurobiologists, and psychologists conducting work on language, memory, and motivation require access to the new generation of relatively noninvasive neuroimaging devices, such as computerized tomography (CT), magnetic resonance (NMR or MRI), and single-photon and positron-emission tomography (SPET and PET). Since these devices are now usually installed in clinical medical settings, gaining access to them by nonclinical or nonmedical researchers is difficult. Yet such devices are far too expensive for nonclinical investigators to acquire and operate strictly for research purposes.

Animal Care

We recommend an increase of $5 million annually to improve facilities for animals used in behavioral and social sciences research.

For researchers using animals, it is essential that funds be available to ensure that the animals receive proper and humane care. Mandatory standards for animal care have risen greatly in recent years. Investigators who use animals in research have generally applauded the change, not only for ethical reasons, but also for practical ones; most research projects benefit from the increased health, longevity, and well-being of their animal subjects. Improvements in facilities for animals are very expensive, however, and little provision has been made to enable researchers to afford the costs of meeting the newly mandated standards.

DATA RESOURCES

A number of large-scale funded data bases serve behavioral and social sciences research. Some of these data bases entail costs comparable to capital pieces of scientific equipment. For example, the 1980 decennial census of population and housing in the United States cost more than $1.1 billion to plan, collect, and process. Of course, most of the cost of the census is to meet constitutional and statutory requirements for information about all of the people and places (not just representative samples) in the United States. The research uses of the decennial census, which are very important, are virtually free riders on this massive exercise. (It should be noted that the research benefits of the decennial census are not captured until the census is published. It took up to 6 years after the 1980 census was taken to publish or release many of the detailed statistical tables and data tapes that are of use to researchers. Planning for the 1990 census should stress issuance of research data within 2 to 4 years.)

A more typical large-scale sample survey designed to test scientific theories or to illuminate certain empirical processes costs far less than $1 billion. The World Fertility Survey, for example, developed data on family size, family planning, and fertility rates in 62 countries at a cost of roughly $50 million during 13 years. The continuing Panel Study on Income Dynamics, covering the changing fortunes since 1968 of (originally about 5,000) American families, costs about $2 million a year to update, analyze, and make available to researchers.

There are three matters of particular importance with regard to data resources. The first matter is the implementation of appropriate criteria in generating new large-scale research data collections, especially longitudinal ones. The second concern is the maintenance and enhancement of research access to governmental microdata files and record systems. A final important set of data resources that need attention are the records of local governments and large corporations.

Large-Scale Data Bases

We recommend a two-track review process for all large-scale data collection proposals, including the $40 million in new collections recommended in this report: a technical review of the scientific significance of any such study, carried out in the same context as investigator-initiated proposals of smaller scale, but without direct reference to budgetary concerns; and, in parallel, an evaluation in terms of the special design features needed, for which there should be distinct review criteria, uniform for all large-scale collections.

One of the trends in the behavioral and social sciences during the past several decades has been a spectacular increase in the size of many data bases used in research. Large-scale longitudinal studies in particular have facilitated major advances in knowledge concerning psychological development (Chapter 2), the political behavior of electorates and legislatures (Chapter 3), and population dynamics (Chapter 4). In the future, new large-scale survey instruments will be needed to promote research advances in two major areas, human development (including cognitive, emotional, social, and health-behavioral factors) and the organization of work. The foreseeable need is for more, larger, and more complex data collections, including the development and maintenance of technical and administrative support facilities.

It is sometimes argued that large-scale data bases are involved in a "big science versus little science" contest, but the human scale of most analytic projects utilizing large-scale data collections is not different from other kinds of projects: it is virtually always a matter of individuals or small groups of close collaborators. The question is not whether one big team will displace many little teams, but how many small analytic teams can rely on any single, large data collection. The relevant competition for resources is really between the researchers who can share a data base and those whose research cannot benefit from such sharing.

Large-scale investments present unusual problems in deciding which projects to support. What mixture of data sets with how much replication among them will best serve researchers in different fields? Under what circumstances is one comprehensive large-scale effort preferable to several smaller projects? When do the particular scientific advantages of panel studies—relative, for example, to repeated cross-sectional surveys—or matched-comparison samples justify the particular kind of long-term commitment attached to them? Under what circumstances can randomized field experiments be effectively coupled with longitudinal surveys? Foremost among the difficulties is that the value of data cannot be assessed separately from an assessment of the questions different researchers wish to answer; the skills and analytic competencies of the principal investigators; and technical matters of optimal sample size, sample population, adequacy of the survey instruments to be used, and other methodological issues.

Prospective or longitudinal panel studies, which collect research data on a substantial sample of respondents over a period of years, provide a particularly valuable basis for testing causal connections—for example, in the development and course of criminal careers (Chapter 2)—that are difficult to find or confirm in short-term or cross-sectional studies. But longitudinal studies cannot be undertaken lightly; they require a long-term commitment, sometimes up to 30 years or more, to ensure that their advantages are fully exploited. Before such a study begins, significant users must conclude that it is of high potential value, a research organization must exist and be committed to keeping the study functioning, and funding sources must be committed to long-term support. Such studies are especially sensitive to interruption. Postponing or ceasing operations for a few years can ruin the fundamental scientific rationale for the existence of a longitudinal study: the systematic and continuous accumulation of useful data over time. Therefore, once such a study is established, it should be discontinued only after determination on the basis of specific criteria and a well-defined review that its maintenance is no longer the best way to commit long-term resources. Two criteria should be key in such a determination: Has the field moved sufficiently beyond the questions the data set was designed to answer? Have sample loss and inadequate replacement led to excessive drift from statistical representativeness?

Behavioral and social scientists have developed a series of criteria to evaluate new and continuing proposals for longitudinal surveys and other large-scale, long-term commitments. One set of criteria concerns the survey's methodological contributions, as discussed in Chapter 5. A second set is the substantive and theoretical importance of the research questions that the study is designed to address. Other criteria derive from seeing how large-scale data collections have fostered the development of more effective research organizations. The most important institutional innovations are new forms of data dissemination and new modes of scientific decision making (or intellectual governance) within the research community.

National and local data libraries designed to disseminate large-scale survey data for research purposes grew rapidly in the 1960s and 1970s. Their initial role was to serve as data archives, creating computerized files of data such as successive public opinion polls, county-level election returns, census information, and congressional roll-call votes. Present archiving initiatives look toward providing effective remote access as well as automated indexing, cataloguing, and search capabilities to aid researchers in identifying data for specific research interests. These facilities also encourage behavioral and social scientists to combine data and conduct temporal and geographical comparisons that go beyond any one data collection.

A major constraint on making archived data easily and quickly available is cost. The development of computer networks and the capacity for remote access is very promising on this score, and special-purpose capacities can be developed where needed (such as the University of Wisconsin SIPP-Access Center, to facilitate research using the Survey of Income and Program Participation conducted by the Bureau of the Census). Access has also been facilitated by requiring grantees and contractors to generate well-documented and readily accessible public-use tapes and by providing resources for them to screen and respond to requests for additional data or documentation. But for the most part, new data collection efforts are likely to depend heavily on existing archives that are proficient in providing documentation, formatting and organizing data files, responding to diverse technical specifications, and using the expertise in data management that is essential for distributing data to researchers in different institutions with different data-processing and computational facilities.

Important as it is, the archive function occurs at a late stage of the data collection process. Consequently, explicit arrangements have begun to be developed for taking potential research users' interests and needs into account much earlier in the design and operation of large-scale longitudinal studies. These arrangements have included appointing a national board of overseers to these projects as trustees for the relevant communities of users. Such boards have been appointed for each of the three largest research data collections that are national in scope and support but conducted outside the federal government: the National Election Studies, the Panel Study of Income Dynamics (both at Michigan's Institute for Social Research), and the General Social Survey (housed at the National Opinion Research Center at the University of Chicago). These boards of overseers have actively sought the participation of many researchers in user conferences, pilot studies, and ongoing design. More significant are the new capacities and responsibilities developed by the overseers, principal investigators, host institutions, and permanent professional research staff in day-to-day charge of these data collection efforts.

Host institutions, while benefiting from identification with major data collections, also accept organizational responsibility for maintaining intellectual leadership, appropriate facilities, and administrative and technical personnel. This responsibility is especially important in the case of longitudinal studies, as principal investigators retire, change career interests, or otherwise reduce their commitments and as significant changes in study content, measurement technique, research design, or collection methods accentuate the importance of staff methodological and technical expertise. Such projects thus engage their host institutions, far more deeply than do most smaller-scale projects, in the health of the research enterprise at a national level.

To meet these national responsibilities, the institutional mechanisms of data archiving and collective intellectual governance of large-scale data collections should be further expanded. First, they should try to encourage simultaneous independent analyses of all data that have a bearing not only on important scientific questions, but also on matters of public policy. Independent inspections, analyses, and criticisms of important new data are needed to inform public debate. The value of these procedures is evident in previous work with surveys of educational achievement, studies of income maintenance schemes, and experiments with alternative patterns of police patrol.

Institutional mechanisms should facilitate collection of data in a manner that permits easy linkage to other data (see below, "Research Access to Government Data"). The maintenance of an ongoing survey should also be viewed as a core activity to which independently designed but fully compatible special-purpose studies may be added on for limited durations, for example, for one or two waves of a longitudinal survey. Such studies could be methodological or substantive (see Chapter 5). Finally, periodic evaluations should be made of data-sharing strategies, accession and acquisition costs, and the scientific and practical results of the data collection. These evaluations should be published: they can serve as a basis not only for evaluating the projects themselves, but also for advances in knowledge about how to evaluate and improve all such large-scale data collections.

Research Access to Government Data

We recommend that federal statistical agencies and programs work toward more access to federal data bases by behavioral and social sciences researchers. Risks of disclosure need to be balanced against the prospective research benefits (including *the practical importance of findings) that may result from access to the data. We note that past research access to many privileged, confidential, and anonymous data sets in the United States has not produced a single case of harm involving a breach of the anonymity, confidence, or privacy of any respondent. We recommend $8 million annually for a series of new initiatives in research access to federal data (discussed in earlier chapters of this report).*

Much research in the behavioral and social sciences benefits from data collected, processed, and made accessible by the federal statistical system. Research on labor markets, social mobility, public finance, criminal justice, and other areas relies very heavily on federal surveys and administrative statistics. These data are not usually collected for the sole or even principal purpose of advancing scientific knowledge, but rather to develop information relevant to the missions of the agencies collecting them. Yet the products of federal data systems are often vital to advancing basic research. In the absence of these statistical records, the sheer effort needed to design and collect this kind of information de novo would drastically increase the direct costs of research.

In many cases, potentially valuable data are not available in fully usable forms. An outstanding instance is decennial census data. The Census Bureau did not begin producing public-use microdata sample files until 1960, and so immensely important information on the massive social changes that had taken place in this country in this century was not available. A large, joint government-academic project involving many university scientists has resulted to date in the production of public-use sample files for the 1900, 1910, 1940, and 1950 censuses. This new data resource greatly increases the informational payoff from these very large national investments in censuses. These data are already proving important to the development of models of population change under varying social and economic circumstances.

The National Science Foundation and other agencies have been prominent in the effort to enhance the research usefulness of federal statistical data and to facilitate access to them. To help make national and international data more adequate scientifically, the National Science Foundation and the Census Bureau are providing support to an American Statistical Association fellowship program aimed at further upgrading the study potential of current Census Bureau data products and making them more directly usable by the research community. Programs of on-site study at federal agencies—of which the American Statistical Association research fellows program is a model—provide an opportunity for researchers to analyze data that may be too expensive to prepare for public release. These programs also increase the opportunities for data-collecting agencies to evaluate the research importance of requests to include particular critical items on repeatedly administered survey instruments.

The federal statistical system, when it is able to serve both agency missions and research needs, is a very efficient resource. Moreover, researchers over the years have contributed substantially to the technical improvement, enrichment, and utility of mission-oriented federal statistics. Yet research access is increasingly jeopardized by concerns about anonymity, confidentiality, and privacy rights—despite the fact that researchers using survey-type data are hardly ever interested in discovering the individual identities of respondents and not at all interested in revealing such identities or using them for nonresearch purposes. Nor have researchers in fact ever been accused of improper disclosure or use.

The most intense concerns about the privacy or confidentiality of records actually refer not to research use but to the routine use by state and federal agencies of each other's automated records for administrative, investigative, and intelligence purposes. These practices were recently reviewed in a report by the Office of Technology Assessment, *Federal Government Information Technology: Electronic Record Systems and Individual Privacy* (1986). This report noted that "the widespread use of computerized data bases, electronic record searches and matches, and computer networking is leading rapidly to the creation of a de facto national data base containing personal information on most Americans" (p. 3). Virtually all the activities reviewed involve access to and study of files that completely enumerate a given population and provide exact identifiers. Research uses, in contrast, involve taking a small proportion of a set of files, generally, a structured random sample, either ignoring or actually deleting from the selected records any individual identifying data before the analysis begins.

Other concerns involve more sophisticated technology. As government agencies and a variety of private companies (such as insurance clearinghouses, direct-mail advertisers, and credit bureaus) amass more extensive and detailed data files, the possibility increases that someone with a large-scale record system in which identities are known could use computer-matching software to identify individual respondents even in a set of anonymous microdata, that is, in files containing individual records from which all identifiers (such as names, addresses, and social security numbers) have been removed. The accumulation of large and detailed microdata files both in government and in business may make the "signatures" of unusual individual records easier to discern through matching techniques.

Federal data-base administrators are becoming very concerned about their own liability in this connection. Agency staff who have been working with researchers for many years perceive increasing pressure to reduce the amount and nature of data they release, despite the fact that the need for these data in research is expanding and despite existing practices that guard against unwanted disclosure. These practices include releasing anonymous data in which geographical location is masked or the minimal identifiable geographic unit is no fewer than 100,000 residents; combining detailed items into composites; truncating information about such items as age or family income into broad categories; and withholding certain details about the sample design. When the problems of identification are especially severe—such as with data concerning very large business firms or very wealthy individuals, because there are relatively few of them—scientists have developed and used the technique of microaggregation for the limited release of confidential data. In microaggregation a series of "average" firms are created from the data, and research files are produced on these average units rather than on actual individual firms. Behavioral and social scientists have been in the forefront of the development of such techniques. Moreover, research users can be placed under legal obligation (and associated sanctions) not to make or permit nonresearch use of protected data to which they may gain authorized access. The liability for any breach then lies with research users, greatly reducing the liability exposure of the agency that is the source of the data.

Despite such safeguards, there have recently been instances of denial of research access to federal statistics that were previously made available in sample form, for example, the Continuous Work History Sample from the Internal Revenue Service. Researchers are concerned that such denials do not become a trend. An overly stringent interpretation of acceptable disclosure risk could drastically decrease present levels of access to some very important data, at a time when austere statistical budgets already threaten established standards for data preparation, publication, and dissemination.

Significant research advances in several areas are likely to depend critically on the availability of large-scale sample microdata bases that can be created by merging administrative records with sample survey results. Such merging of records on a sample basis for research use has encouraging precedents. For example, an earlier effort linked a sample of tax files from the Internal Revenue Service, data from the Current Population Survey, and selected Social Security files for the year 1973 and rendered them suitable, including suitably anonymous, for research. A good candidate for producing new research advances is the proposed linkage of Social Security and tax data to information from the nearly 30,000 households sampled in the longitudinal Survey of Income and Program Participation now being conducted by the Census Bureau; such linkage would provide an unparalleled resource for studying the microdynamics of household economic behavior. Similarly, the availability of geographically disaggregated microdata on individuals, firms, and households would immeasurably benefit research on migration, regional growth, and urban ecology.

This discussion has focused on access to quantitative survey-type data, but there are other critically important types of governmental data that, for reasons of confidentiality, security concerns, and lack of funding, have not been developed as research resources. Development of some or better research access to such data—whose costs of collection have been and will continue to be covered by their mission uses—is among the most cost-effective research expenditures that could be made in any area of behavioral and social sciences.

Corporate and Local Government Archives

We recommend $2 million a year for new studies to determine the research value, methodological problems, and costs of preserving and gaining improved access to major private record centers and unused corporate and local government archives for research purposes.

In contrast to most federal agencies, many organizations that collect and store potentially useful data on social and behavioral processes have no history of making them available for research. Principal examples here are most of the files of major corporations and the archives of local governments; in most cases, the data are paper files that are stored in warehouses.

There are three stages in gaining access to such data. The first is to keep corporations and local governments from destroying or throwing away their records without regard to their potential value for research. The second is to gain permission from their custodians to study them. The third is to place such records or samples of them, as appropriate, into computer-readable form and train technical staff or develop appropriate documentation to assist potential users. In some cases these stages may involve significant costs, and the relation between potential research benefits and costs of preservation, availability, and conversion must be carefully considered. In many instances the benefits may outweigh the costs, but detailed plans and criteria for archival practices have not been developed.

To increase access to research-relevant data for basic studies as well as to evaluate federal programs, nongovernmental archives with large record systems should be considered for receipt of research funds. One desirable way to increase the usefulness of data archives—public and private—would be to provide them with trained staff or statistical analysis. Capacities for producing properly protected data files, or performing statistical analyses when it is impossible or too costly to protect the files adequately for release to researchers, would improve the range of potential resources at relatively low cost. Multivariate analysis often requires little more than covariance matrices for a sample and for some important subpopulations, a level of aggregation that surely protects the privacy of individual records. The data held by Blue Cross/Blue Shield and other carriers of medical insurance, major automobile and life insurance companies, and the like could be made more accessible to research uses if each major record center had staff with research-oriented statistical and programming expertise. With such capacities, data archives could then respond to requests for analysis of their holdings either by releasing appropriately protected data or by performing some (or all) of the statistical analyses requested by the researcher, who could be directly charged for some (or all) of the costs involved.

FUNDING RESOURCES

In light of the unpredictability of research, there is inevitably a need for patience and adaptability on the part of those who produce scientific work and those who underwrite its costs. The ideal is to maintain a funding system that is fiscally responsible, flexible, and responsive to new developments. It should encourage competition among new ideas, sustain productive investments, and close off research avenues that are no longer productive. It is impressive that present funding arrangements manifest most of these ideal components most of the time. The changes needed and recommended in this chapter, while important, should be regarded as improvements at the margin of a basically sound system of support.

Probably the most vexing problem facing researchers and the agencies that fund them is how to decide between funding individual scientists to conduct highly specific, time-limited projects and providing support for more extended, less specific research designs and structures. The latter include arrangements for shared access to expensive technologically advanced equipment, groups of investigators working together on difficult multidisciplinary problems, and longitudinal studies. Such facilities and organizations are beyond the scope of individual investigation and require making choices that can have far-reaching consequences and be difficult to reverse. The development of a complex data base or the creation of a research center may well constitute valuable enterprises that can contribute to the scientific work of many investigators; but when a large-scale project may command, say, an annual sum equal to one-third of the federal budget lines presently devoted to basic research in one of the core disciplines, a serious debate is joined.

Modes of Support

We recommend that the major mechanism for supporting fundamental research in the behavioral and social sciences continue to be individual grants awarded under a scheme of competition among intellectually similar projects, with evaluation con-ducted by scientifically qualified and organizationally disinterested individuals. We recommend an aggregate increase of $70 million annually in the level of support for such research (as detailed in previous chapters of this report).

The mainstay of research support in the behavioral and social sciences is the modestly sized (roughly, $30,000–$70,000 per year), short-term (up to 3 years), competitively awarded grant, administered through a standing organization (often a university), for the part-time support of research by an individual investigator with one or more assistants who are often graduate students work-ing part-time on research. Such grants provide resources to carry out the study specified in the investigator's proposal, with the requirement that technical and financial reports be submitted periodically to the funding agency. The research results are expected to be written up and published in the scientific literature.

Most scientists believe that such grants are a very good way to nourish the highest quality and productivity of research. The investigator-initiated grant process is a system of repeated direct competition for new funds among discrete proposals. In each proposal, the prospective investigator discusses his or her past research accomplishments, details the methods he or she proposes to use and the theories to be investigated, and estimates the significance and promise of the proposed work. Evaluation of these proposals is generally carried out (often on an anonymous or confidential basis to ensure candor) by other researchers who possess the scientific expertise needed to understand and appraise the work. Evaluators are, as a rule, precluded from judging the merits of any proposal if they are closely involved with the particular applicant or the investigator's institution. In most agencies, including the National Science Foundation, the National Institutes of Health, and the Alcohol, Drug Abuse, and Mental Health Administration, the technical evaluation is conducted by an external scientific review panel supplemented by ad hoc consultants. In other agencies, including the U.S. Department of Defense, the scientific evaluation is internal (in-house), and the system is usually subject to a general periodic appraisal of the program by outside evaluators. There are also a variety of mixed systems, for example, in the U.S. Department of Education.

It is essential to preserve what the scientific community widely agrees is especially valuable in the current system of allocating research support.

Grant Size and Duration

We recommend that funding for individual research grants in behavioral and social sciences be revised. Each proposal and award should include adequate funding to purchase appropriate research materials, contact subjects or acquire and maintain research animals, collaborate and communicate with geographically dispersed colleagues, acquire or maintain and fully analyze research data, and support research staff, including the principal investigator(s). The proposals that receive the most favorable evaluations in competitive review should be fully funded even if the total number of grants awarded must be constrained in order to do so. In addition, funding agencies should increase the present average durations of awards in the behavioral and social sciences in order to secure continuity of research and reduce the amount of time spent on preparing and reviewing proposals.

Much research in the behavioral and social sciences is both labor intensive and equipment intensive. For example, further advances in the knowledge of how fundamental mental capacities develop will depend on using video devices that can record where infants look or place their hands and computers programmed to track and analyze facial movement. Research of this sort also requires staff to arrange with parents for visits to laboratories and to conduct interviews to obtain other relevant social, behavioral, and physiological data. Similarly, without considerable investment in both personnel and equipment, it is virtually impossible to run a modern laboratory studying the neurophysiology of memory, the conditioning of complex primate behavior, or the variations in simulated commodity trading under varying prices and market arrangements. A field research facility to study fertility determinants across a larger area's population requires a substantial staff to develop a valid and reliable sample, track the respondents, and periodically conduct interviews about births, deaths, marriages, illnesses, family finances, and intimate beliefs and practices over a lifetime. Such projects also need substantial space to house staff, records, and computational equipment.

We stress these facts about personnel, facilities, and the resulting expenses of research, because, all too often, funding sources or officials who want the results of research or even universities still think that state-of-the-art work in the behavioral and social sciences requires only a principal investigator, one or two graduate students, and a slightly larger office allocation. The size of the typical research grant in the behavioral and social sciences often precludes substantial use of modern equipment or hiring trained staff, including the technicians, predoctoral assistants, and support staff, needed to carry out high-quality research. Because of year-by-year administrative decisions made by staff and review panels in key federal agencies such as the National Science Foundation, the trend of decreasing budgets experienced over the past decade has not led to appreciably fewer proposals being funded, but to reductions in the size of each grant in many research programs (see Appendix A).

Another response to budgetary stringency by granting agencies has been to decrease the duration of grants. For example, although a maximum duration of 5 years per competitively reviewed proposal is possible at the National Science Foundation, the average standard (fixed-term) grant is typically 1 to 2 years. Continuing grants, which can continue for up to 5 years, comprise only one-third of all National Science Foundation awards in the behavioral and social sciences. In the National Institutes of Health, which has not been constrained in the same way, investigators and review panels have nevertheless been oriented to short-term proposals and awards even when an investigator's proposal is on a more extended timetable. This pattern of short-term funding significantly increases the number of applications that investigators must prepare and reviewers must evaluate. As a result, significantly less time is available for research.

If the costs of doing state-of-the-art work have come to exceed the size and duration of the typical current grant, funding programs must resolve a difficult dilemma: whether to continue with too-small awards across the board, which may inhibit or cut the scope of new projects, or to constrain the number of grants funded in order to foster highest quality work. The best solution would be more *and* larger grants, but at this time growth in size should take priority over growth in number.

The Disciplines and Interdisciplinary Research

We recommend that procedures for evaluating and funding interdisciplinary research proposals be reexamined. Most of the reviewers of those proposals should have scientific interests and competence in interdisciplinary areas. Staffing needs to be adequate to permit appropriate handling of interdisciplinary projects, which often require greater time and attention. We further recommend that the National Institutes of Health and the Alcohol, Drug Abuse, and Mental Health Administration reconsider the criteria they use to evaluate behavioral and social sciences research. Support for the study of behavioral and social phenomena should not be restricted to factors specific to particular diseases, but should extend to factors appearing important in the etiology and treatment of a number of diseases. The National Science Foundation should reconsider its programs for supporting studies of intelligent systems—natural ones (humans and animals) and artificial ones (computers)—which were separated after a promising start at developing an interdisciplinary program.

Most research in the behavioral and social sciences (as in most science) is carried out within the confines of familiar academic disciplines. The largest behavioral and social sciences discipline, in terms of personnel trained, is psychology, followed by economics, sociology, political science, and then the substantially smaller disciplines of anthropology, linguistics, and geography. In addition, many scholars whose degrees or primary affiliations are in the fields of education, history, law, management, operations research, philosophy, police science, psychiatry, public health, and statistics identify with the behavioral and social sciences on the basis of their research interests, methods, or theories. Researchers oriented to the behavioral and social sciences are sufficiently numerous and mutually cognizant in some of these nearby disciplines to have created specialized research societies. The discipline of statistics, which is practiced, taught, and contributed to by scientists from a variety of backgrounds, is often considered and treated as fundamentally a part of the behavioral and social sciences.*

The behavioral and social sciences disciplines are durable structures for professional certification, advancement, communication, and academic instruction. Research emphases, theories, and methods come and go, but uni-

---

*Descriptions of the main substantiative concerns of the respective disciplines can be found in Chapter 2 of an earlier report of this committee, *Behavioral and Social Sciences Research: A National Resource*. Washington, D.C.: National Academy Press (1982).

versity departments and disciplinary associations continue. Advisory and honorary groups such as the National Academy of Sciences and the American Academy of Arts and Sciences categorize their memberships along standard disciplinary lines. However, many frontier areas of research sprawl across disciplinary boundaries and have hence to be considered interdisciplinary. For example, many researchers who are studying the development of human expertise are as much computer scientists as they are psychologists. Research on addictive behaviors combines biological and behavioral principles in developing new theories and research designs. The comparative analysis of languages and religions and the study of contemporary international structures and processes draw on the contributions of anthropologists, economists, political scientists, sociologists, and historians.

Another feature that contributes to interdisciplinary work is the transfer of methods and theoretical approaches. For example, formal models originally developed in logic, mathematics, and computer science now pervade many other disciplines. Formal mathematical or philosophical training has come to be common among those who do research on the nature and development of spatial, mathematical, and logical thought. Developments in measurement theory have found applications in marketing, the study of risk and utility, and theories of sensory processes. A number of general mathematical topics, including stochastic models, systems of linear and nonlinear equations, and other geometric and algebraic structures, have found applications in econometrics, models of conflict resolution, and the sociology of occupational attainment.

Over time, interdisciplinary research can be best seen as an interaction that develops across and works back into the disciplinary structures. Because research frontiers often involve problem definition, exploratory knowledge, development of new methods, or strong policy interest, creative scientists from various disciplines can often enter them relatively easily. Progress in such research usually accelerates when the problems are sufficiently defined to enable the application of the most advanced technical tools from one or more disciplines—tools that are constantly being improved and refined. As progress in research crystallizes into new bodies of accepted knowledge, continued work in the area may become incorporated into the core curriculum (and other durable structure) of one or more disciplines—in some cases directly displacing other content and transforming the disciplines—or may become the basis of a newly recognized specialty. Over time, such a specialty may even approach or achieve the status of a discipline. Examples are the emergence of linguistics, mainly from anthropology and comparative languages, and of criminology, from sociology, law, and other fields.

These kinds of developments—which count significantly among the research frontiers highlighted in this report—have stimulated seed grants from private foundations, interest from time to time on the part of federal agencies, and occasional support from colleges and universities. But they go against the grain of standard procedure. For example, consider a collaborative study of language learning that brings together linguists and philosophers located in a humanities division; neurologists in a school of medicine; psychologists who, depending on the institution, may be in either social, biological, or physical sciences divisions; and computer scientists and applied mathematicians in schools of engineering. Developing a collaborative proposal means dealing with different assumptions about normal teaching loads, different mechanisms of support for graduate students, and even different academic timetables.

If such an interdisciplinary group applies to a diversified funding agency such as the National Science Foundation, the proposal is likely to be of partial interest to different divisions within a directorate and even to different directorates. Unless the agency staff has sufficient reserves of time and energy to adopt such a proposal and give it special handling, shepherding it diligently through all the possible sources of support, the review system is likely to treat it more as an unworkable processing problem than an unusual scientific opportunity. When agency staff are reduced or their grant loads are increased in the interest of greater processing efficiency, the capacity to respond creatively and effectively to such opportunities diminishes and, ultimately, disappears. Unless funding agencies—and universities—periodically consider whether they have the capacity to facilitate good research that does not fit the established patterns, they are not likely to have such work in progress, and new interdisciplinary research will be inhibited.

The most challenging aspect of interdisciplinary research is to assess whether new developments are genuinely powerful or are passing fads. The general issues in evaluating interdisciplinary projects are, first, to determine in which budgetary pool a proposal competes and, second, to decide the intellectual criteria on which it is to be judged. New efforts are needed to consider those issues. If review is primarily by disciplinary panels, they must include researchers who are experienced in interdisciplinary projects. Alternatively, agencies could constitute a specialized panel or panels to evaluate all interdisciplinary proposals as a group, with details of program-based funding to be worked out separately and subsequently. The most radical step would be to generate a separate resource pool for innovative interdisciplinary projects, which would not compete on a case-by-case basis with disciplinary proposals.

Two recent examples of lack of receptivity to interdisciplinary developments are of particular concern. First, in recent years two major research support agencies in the U.S. Department of Health and Human Services, the National Institutes of Health and the Alcohol, Drug Abuse, and Mental Health Administration, have increasingly stressed that behavioral and social research should directly address specific disease syndromes. This emphasis, reinforced by the sizable representation of clinically oriented physicians on the behavioral research review panels of these agencies, excessively narrows their investments in many useful areas of basic research. A focus on interdisciplinary areas of etiology, prevention, treatment, and services research would alleviate this unfortunate tendency and strengthen the research portfolios of these agencies.

The second specific concern is the former Division of Information Science and Technology at the National Science Foundation, which in 1986 was transferred from the Directorate for Biological, Behavioral, and Social Sciences, where it had drawn information science research into close contact with research on cognition, language, and other areas of the behavioral and social sciences. It was transferred to the new Directorate for Computer and Information Science and Engineering and reorganized as a Division of Information, Robotic, and Intelligent Systems. As a result of the move, the unit has ceased to fulfill its important role in supporting basic behavioral and social sciences research projects with strong information science components, to the detriment of both areas and especially of their interdisciplinary intersection. It would be substantially more productive in our view for the National Science Foundation to redirect this program, to bring studies of artificial information processing systems back into contact with studies of natural ones.

Interdisciplinary Research Centers

We recommend that private and federal funding agencies encourage submission of proposals for relatively long-term multidisciplinary research activities focused in research centers, housed either in single institutions or in consortia. Funding should be provided for new centers as well as for centers of proven effectiveness. Review of center proposals should be at least partly independent of single-investigator proposals. A program to provide development or pilot funds for center proposals is necessary in nearly all behavioral and social sciences fields. We recommend that research centers receive approximately $25 million in new funding.

Interdisciplinary research centers are intended to solve two closely related problems. One is the need to bring together a critical mass of scientists for research that by its nature requires interdisciplinary input. The other is to provide an appropriate physical, logistical, and administrative setting for advancing interdisciplinary research that may not be limited to a particular focus or set of problems. There are several examples of national and international centers and institutes that have endured for decades and have contributed to significant advances in the behavioral and social sciences.

Many of the same basic questions that were discussed in connection with evaluating new large-scale data collections apply to new research centers: How worthwhile scientifically are the anticipated results? What fraction of the new funding of the relevant fields should go into a center or centers? Can the same results be achieved as efficiently by encouraging new directions of research through individual grants or already established centers? Additional considerations are the frequency and nature of evaluation of the work of a center; the ways in which less successful centers might be phased out; and the institutional locus (for example, in a university or an independent institute).

Most initiatives for centers discussed in Chapters 1–5 are oriented either toward the efficient use of a multiuser laboratory or toward a large-scale data collection project. In such cases, the sheer size of the laboratory or survey project generates the need for a centralized unit to facilitate the work. In some other cases, a center is seen principally as a mechanism to ensure interdisciplinary communication. For example:

- **Learning and memory.** Centers in this area could bring together investigators in psychobiology, neurobiology, cognitive science, mathematical neural modeling, and the experimental analysis of behavior to focus on particular memory circuits, networks, or aspects of learning and memory. A particular advantage of such arrangements is to facilitate the sharing of equipment, such as imaging devices. One or more centers could be devoted to studies using animal subjects, which would introduce economies of scale in financing improvements in the conditions under which the animals are maintained and studied. One or more research centers could focus on particular aspects of clinical learning and memory disorders, for example, aging and amnesia.

- **Affect and motivation.** Centers in this area could bring together researchers among whom there is at present too little sustained communication, including psychiatrists and neurologists who are concerned with research on and treatment of affective and motivational disorders and psychologists and sociologists who are engaged in research on the interpersonal and social factors that shape and condition emotional development and expression.

- **Demographic behavior.** There is substantial warrant for a new demographic research facility in the developing world, particularly in Africa, to engage in epidemiological, ethnographic, and demographic work similar to that now carried out by the International Centre for Diarrheal Disease Research in Bangladesh. Data are needed on biological characteristics, patterns of individual decision making regarding marriage, childbearing, and childrearing, and institutional and cultural features in other regions, such as Africa. Sources used would include demographic records and surveys, field observations, and historical records. Such a center would enable researchers to track discrete cohorts of individuals through several decades of life and would be a stable foundation for research on childbearing, mortality, morbidity, population, and the institutional and cultural factors that affect these phenomena among hundreds of millions of people about whom very little is known.

In general, the funding of such research centers—but not a number of other large-scale activities such as longitudinal studies—can be arranged on a two-tier basis. One tier is a core grant designed to provide the autonomous infrastructure and flexibility needed by the center. A core grant covers basic operating and administrative costs, some shared laboratory or computing facilities, common staff members, and a certain amount of discretionary research funding mainly for new investigators before they move to the second tier. That second tier consists of proposals submitted by individual investigators and reviewed in competition with all other investigator-initiated grants. This second tier, of course, forms the backbone of the research endeavor of a center. The advantage of such a partition of funding is that research progress will be regularly and appropriately monitored by the relevant specialists.

THE PROBLEM OF VOICE

The place and role of the behavioral and social sciences in the administrative arrangements of relevant federal agencies should be critically reappraised. It is necessary to ensure continuous high-level representation of the scientific needs and opportunities in these fields. The coordination of these sciences and their general advisory roles should be strengthened by establishing a mechanism to coordinate interagency policy on behavioral and social sciences research.

One feature of the present situation that works against the best interests and the best utilization of the behavioral and social sciences is the way they are situated within the administrative structures of federal research agencies. If one looks to research-oriented academic institutions, it is customary for faculty members to be affiliated with divisions of humanities, physical sciences, life sciences, and social sciences or schools of business, law, medicine, and other professions. (History is located sometimes with the social sciences and sometimes with the humanities, and psychology is located sometimes with the life sciences and sometimes with the social sciences.) This division permits behavioral and social scientists to negotiate effectively the immediate administrative environments in which they work and, in most universities, to gain reasonable access to (and occasionally to find colleagues in) high administrative positions.

If one looks at the behavioral and social sciences in federal research agencies, there is no parallel to academia. For example, the National Science Foundation has its Directorate for Biological, Behavioral, and Social Sciences. The title of the directorate, except for its nonalphabetic sequence, suggests perhaps a certain degree of parity, but that is not so: the budget for the behavioral and social sciences was about $50 million in fiscal 1987, compared with $200 million for biological research. Moreover, since its inception in 1974, this directorate has been headed by biologists. There are similar or even more disproportionate relationships between biomedical and behavioral and social sciences research in every institute in the Alcohol, Drug Abuse, and Mental Health Administration and National Institutes of Health.

This kind of subsumption under related fields has led in the past—and could lead at some future date—to misperceptions at the highest policy levels about the scientific opportunities and needs of the behavioral and social sciences. This problem of misperception is aggravated by the paucity of high-level co-ordination between agencies of the federal government that fund behavioral and social sciences research, including the Departments of Health and Human Services, Defense, Labor, Education, Justice, Commerce, State, and Housing and Urban Development, and independent agencies, especially the National Science Foundation and the Smithsonian Institution. To deploy scarce resources wisely and effectively, there should be efforts to develop complementary (not, of course, monolithic) research and funding policy for the behavioral and social sciences across the federal government.

7

Raising the Scientific Yield

7

Raising the Scientific Yield

The behavioral and social sciences have made many notable advances in the nearly two decades since the completion of *The Behavioral and Social Sciences: Outlook and Needs* (1969), the last survey similar to this one. Looking forward, the scientific opportunities are diverse, intellectually inviting, and profound in their implications for the further understanding of individual and social behavior. In this summary chapter we review the thematic highlights and salient research advances covered in previous chapters and the new investments and modifications in research infrastructures that are needed for further progress. We recommend that the initiatives and resources we call for be implemented within 3 to 4 years.

RESEARCH FRONTIERS

Behavior, Mind, and Brain

Three major developments have propelled new inquiries into the connections among behavior, mind, and brain: improved observations on the course of individual human growth, exploration of the relationship between the information-processing capabilities of humans and machines, and further discoveries of biological and behavioral commonalities between humans and other animals.

Several major research questions have begun to be answered, and further work is promising:

■ By what process do humans and some animals achieve their advanced and complex powers of visual and auditory discrimination, far greater and more subtle than can yet be attained by technology? Neurophysiological and behavioral experiments and computer simulations have substantially advanced understanding of how the brain analyzes visual data, recognizes color constancy, infers depth and motion, and organizes auditory stimuli over time.

■ How are memories coded, organized, stored, and retrieved? Scientists working in a variety of disciplines have identified distinctive types of memories and genetic limitations on memory and isolated cellular and molecular mechanisms in the brain's transmitting systems.

■ How do human beings acquire knowledge, organize it, use it in reasoning, and implement it in behavior? Experimental studies have uncovered development of complex cognitive capacities in infants. New theoretical principles are being explored that govern how people categorize information, use visual imagery, make decisions, and deal with uncertain situations.

■ How do humans acquire and use that most distinctive of human endowments, language? Neurophysiological studies of the brain, investigations of sign language, and comparative studies of grammatical structures have yielded rich findings on the genetic foundations of language, the association of language functions with particular brain areas, and the formal properties of language. New advances are also evident in the computer recognition and production of language.

Research in some of these areas has come to demand very complex instrumentation for simulating, modeling, recording, and analyzing data in laboratory experimentation. We regard it as critically important to augment the computational and laboratory infrastructure underlying this research. Another distinctive characteristic is the multidisciplinary character of advanced work in these areas, which can involve neuroscience, physiology, psychology, child development, biophysics, biochemistry, ethology, linguistics, statistics, economics, and computer science. Opportunities to build on this call for various support mechanisms, such as new centers of research, interdisciplinary graduate and postdoctoral training, advanced study workshops, and longitudinal studies of cognitive and educational development. In addition, the base of investigator-initiated grant support must be increased to continue this research.

Motivational and Social Contexts of Behavior

Social challenges and practical urgency have reinforced the already substantial scientific interest in affective and motivational states and processes, violent crime, linkages among physical health, behavior, and social contexts, and the nature of social interaction. There are now improved research capabilities to respond to these interests, particularly in the form of better longitudinal study designs and analytic techniques, new observational and assessment technologies, and theoretical innovations.

We note several areas of particular interest, ferment, and advance concerning affective and social contexts of behavior:

■ How have emotional and motivational processes driven human development, human behavior, and human evolution? Advances have been made in pinpointing the neural bases of such motivations as hunger and sex, as well as predatory and defensive fighting. New measures of facial expressions have improved understanding of emotions. The subtle interplay among physiological, neural, psychological, and sociocultural forces that determine the dynamics of eating behavior is beginning to be grasped.

■ How do behavioral and social forces affect the health of individuals and groups? In a knowledge shift as revolutionary as it is by now commonplace, the conventional stress on the biological aspects of human health has been balanced by understanding that behavioral and social forces are major etiological factors in health and disease. For substance abuse in particular, the influence of peer groups, life-cycle changes, media, and market factors has been more fully detailed. The role of stress and risky behavior in the genesis of health disorders has been verified and some of its dimensions pinpointed.

■ What are the determinants of criminal behavior? New and more comprehensive measures of crime rates have been devised. Theoretical shifts are under way, such as moving from studying aggregate rates and their social correlates to studying microdata on individual criminals and the life cycles of persistent and other types of criminals. Longitudinal studies of sample groups with different criminal histories are regarded as an especially effective way to sort out family patterns, peer influences, and law-enforcement policies as determinants of criminal behavior.

■ What is the interaction between personal characteristics and group effects as influences of behavior? Some research has focused on the phenomenon of the self-fulfilling prophecy: if others attribute certain characteristics to a person or group, that attribution can play a substantial role in generating those very characteristics. Other experiments have thrown new light on the effect of group size on task performance and the effect of majority or unanimity rules on the outcome of group decisions.

These lines of research now appear to call especially for new kinds of data sets, particularly longitudinal ones, since many of the phenomena in question—

criminal behavior, substance abuse, the development of affect and motivation—require study over time. New kinds of laboratory and measurement equipment, particularly in the study of motivational contexts, also require substantial additional support. Increases in the amount of investigator-initiated research are needed to sustain and advance quality research, as are additional resources for doctoral and postdoctoral support, interdisciplinary programs of research, and advanced technical workshops and seminars.

Choice and Allocation

Research on the mechanisms of choice and allocation ranges over politics, organizational relations, and economic phenomena. Among the last, the study of markets has played a dominant part, but attention has increasingly turned to clarifying the operation of nonmarket phenomena in certain sectors of Western economies (such as public goods and environmental protection), in the mixed economies of many developing countries, and in Soviet-type economies with central planning.

There are a number of exciting developments and promising strands of research in these areas:

- How much do voting procedures affect the outcomes of votes? At the microlevel, research has demonstrated that an outcome is strongly determined by the order in which a decision-making agenda is arranged, especially if a committee or legislature faces complicated options. More generally, a mixture of theoretical, experimental, and historical analyses are beginning to give powerful assessments of the outcomes and efficiency of alternative voting arrangements.

- How useful are traditional economic assumptions about markets—complete information, rationality of actors, negligible transaction costs? Much recent empirical work has challenged these assumptions and is producing a wider and more realistic range of models of market behavior.

- Under what conditions will actors in fluid or in highly structured marketing situations strike bargains and live up to them? Research on market contracts and bargains has advanced dramatically, and the new methods and theories are being applied to bargaining and its breakdowns in litigation, war, and strike arbitration.

- How well do job markets function, and what are the sources of continuing unemployment in the rapidly changing context of the present U.S. and world economies? New studies of cyclical unemployment are focusing on the effects of implicit and explicit contracts and the resulting unresponsiveness of wages and prices to market conditions. Studies of frictional unemployment are focusing on the processes of job search and job matching. Studies of structural unemployment are dissecting the effects of job segregation (especially between men and women), internal labor markets, and labor migration among firms and regions of the country.

In these areas of research, much valuable work is theoretical in character, and its testing involves applications of models and statistical analyses to data that are readily available in the form of recorded market transactions and economic time series. Consequently, the investigator-initiated mode of research should be given the highest priority for expansion in these areas, with a special eye to the average size and duration of grants. At the same time, the use of panel and longitudinal data, especially on work histories and organizational strategies, is increasing in importance and value and therefore demands increased support. There is also more extensive use of newly devised laboratory experiments for studying the impact of contractual and market rules, behavior under uncertainty, and other topics; this development calls for more appropriate laboratory facilities. While instrumentation needs are probably not as extensive for these research areas as for others, there is great need for upgrading computational hardware and for software development. And as more interdisciplinary work among economists, political scientists, organizational scientists, and psychologists develops, more central facilities and programs for interdisciplinary work should be supported.

Institutions and Cultures

Comparative and historical (including prehistorical) study of the institutional and cultural origins of entire societies has moved forward on a variety of fronts. Researchers have been especially attuned to the interplay between large-scale and local development, and, more generally, between the macroscopic and microscopic levels of social and cultural organization.

Several major questions have yielded to answers, which in turn point to more detailed work:

- What are the evolutionary bases of human social bonding and the formation of extensive human societies? The prehistoric development of families and larger social groups has been highlighted with special attention to the roles of food, foraging from home bases, and particular uses of tools. Hints are also emerging as to the heretofore unknown evolutionary role of language.

- What are the special social and cultural determinants that influence the class of events—births, population movements, marriage and divorce, aging, and death—that comprise demography? There have been substantial new research developments in measurement and theoretical models of demographic change, especially of fertility and migration.

- What factor can best explain those institutional changes that have given rise in the West to what is called the modern world—changes that are now fermenting elsewhere? Research on the history and contemporary status of the family and on the dynamics of religious institutions is providing important and unanticipated new insights into this broad and highly charged topic.

■ What complex of institutional and cultural factors encourages the rise of science and its far-reaching technological applications in society? Historical, sociological, and other research on these questions is moving rapidly as global cases and comparative research opportunities multiply. A much clearer picture of the place of science, including the behavioral and social sciences, in the larger society is also emerging.

■ What are the ramifications of the increasing internationalization of the world? Two growing lines of research are receiving attention: increasing economic, financial, political, and cultural interdependence; and efforts to understand and untangle the analytic complexities and policy implications of international conflict and security.

These areas call for special attention to collaborative group research, in part because much of it is interdisciplinary, but more importantly because it is so often international in character. There are currently very restraining limitations on the conduct of international-collaborative research and very serious problems of sustaining research access to field situations in many parts of the world. These areas also need new resource investments in developing new data bases and in improving access to governmental, financial, and business data. There is a need for major new international and collaborative research centers and a significant expansion of investigator-initiated research grants to make use of archival and other facilities. Expanded support is also needed for the technological base of research, especially computer equipment and software, as well as for graduate and postdoctoral training and fellowships.

Methods of Data Collection, Representation, and Analysis

Methodological advances play a special role in the generation of behavioral and social sciences knowledge. Sometimes these advances arise in the struggle to solve difficult substantive research problems; sometimes they involve separate discoveries; sometimes they involve selective borrowing of techniques from elsewhere; and sometimes they arise as new and better sources of data permit new and more sophisticated techniques to be tried.

Several kinds of methodological innovation are the subject of intense current work:

■ Improvements are occurring in the sophistication and scope of application of four basic empirical methods in the behavioral and social sciences: laboratory experiments, field surveys, ethnographic investigations, and comparative studies.

- Refined techniques are emerging for the measurement of both large-scale and small-scale behavioral and social phenomena, leading to the reduction of many types of measurement error.

- Researchers and theorists are inventing new ways of representing empirical phenomena in symbols and calculations, including log-linear models for categorical variables, multi-item measurement, scaling, clustering, and network models.

- There are advances in statistical inference and analysis, including new techniques of inferring causality, handling multiple parameters that are interrelated with one another, and making estimates in cases of partial data. Advances in computing techniques are closely related.

We recommend new support for purely methodological research, although on a more modest scale than in the substantive areas identified. Several mechanisms are of central importance: investigator-initiated grants, which are the locus of many methodological innovations; summer institutes, colloquia, seminars, and postdoctoral study, where scientists can be exposed to methodological developments in their specialties and to the methods appropriate to their new lines of research; and more powerful computational resources, which are key to further methodological advances.

RECOMMENDED NEW RESOURCES

The character of research covered in this report is varied, but all scientific work retains much in common. In looking across the fields of research covered in Chapters 1 through 5—behavior, mind, and brain; motivational and social contexts of behavior; choice and allocation; institutions and cultures; methods of data collection, representation, and analysis—we find consistent need for human, technological, data, and other resources. This section summarizes the new funds and new procedures that we recommend to ensure the continued growth of knowledge in the behavioral and social sciences. Our overall recommendations for new research initiatives in terms of the various categorical expenditures are summarized in Table 7-1.

Human Resources

Under human resources, we note the critical need to increase the availability of predoctoral and postdoctoral research fellowships in order to encourage the retention of talented scientists in careers oriented to research. At the predoctoral level, we recommend an additional $10 million annually in graduate support, which would cover stipends and institutional costs for about 500 graduate students. The purpose is not to increase the number of students enrolled in

TABLE 7-1   New Research Initiatives—Additional Resources Needed ($ million)

Type of Resources	Resources Needed (by chapter)					
	Behavior, Mind, and Brain	Contexts of Behavior	Choice and Allocation	Institutions and Cultures	Methods	Total
Human resources						**44**
Predoctoral support	2	2	1	2	3	10
Postdoctoral fellowships	3	3	5	6	1	18
Advanced training institutes	1	3	—	1	2	7
Research workshops	1	1	3	3	1	9
Technological resources						**51**
Neuroimaging devices	2	2	—	—	—	4
Animal care	4	1	—	—	—	5
Laboratory technology	12	5	2	1	—	20
Computers and software	7	2	4	5	4	22
Data resources						**50**
Access to federal data	—	—	4	4	—	8
Access to other data	—	—	1	1	—	2
New data collections	5	15	12	8	—	40
Research centers	4	9	4	7	1	**25**
Investigator-initiated grants	20	13	20	13	4	**70**
Total	61	56	56	51	16	**240**

degree programs, but to direct the best students' efforts much more intensively toward research training.

At the postdoctoral level, we recommend an additional $18 million annually in support to be directed to all levels, from new PhDs to senior career awards. We estimate that this support would add approximately 400 full-time doctoral-level scientists to the numbers working at the frontiers of behavioral and social sciences research, which would be a major element in activating many of the other resources recommended here. This recommendation applies particularly to research in the areas of Chapters 3 and 4, which were most heavily affected by reductions in federal support after 1978. We also recommend that awards to potential fellows be made early in the academic year to encourage viewing them as prizes rather than substitutes for professional appointments.

The quality of research can be readily improved by added resources to advanced training institutes for active investigators. Such institutes enable methodological and technical innovations to diffuse rapidly across geographical and disciplinary boundaries. In addition, the cultivation of new theoretical ideas and the coordination of research efforts in rapidly moving specialties can be greatly facilitated by a program of research workshops bringing together core groups of researchers at least once a year for intensive exchange, review, and collaboration. We recommend that $7 million and $9 million, respectively, be allocated for these activities. Experience dictates the need for some or all of these funds to be especially segregated for these purposes.

Technological Resources

Technological resources available to most researchers in the behavioral and social sciences have lagged badly behind needs except at a few sites. The lag has been particularly acute with respect to neuroimaging and the more diversified and specialized laboratory equipment used in research on behavior, as well as the recent upgrading of standards for animal care. We recommend that a total of $29 million annually in new funds be allocated for these needs, concentrated in the research areas discussed in Chapters 1 and 2. The overall requirement for more capable computers and for the development of specialized software (as well as the acquisition of more sophisticated general-purpose software systems) for research in the behavioral and social sciences calls for an additional $22 million annually.

Data Resources

The role of extended data collections in advancing knowledge is a significant feature of behavioral and social sciences research. Such collections are most often acquired and maintained under research auspices and protocols, but a large contribution has been made by researchers using public and private statistical files and historical archives generated primarily or originally for administrative or other nonscientific purposes. The potential benefit of such resources to advance scientific knowledge is very large relative to the incremental cost of converting them into usable research data. We recommend that new funds of $8 million annually be devoted specifically to upgrading and expanding the research utility of federal data and an additional $2 million be devoted to expanding access to corporate and local government files.

We have recommended six new initiatives for data collection in the United States and a seventh initiative involving international efforts. Four of these initiatives cover human development from childhood through middle and mature adulthood, particularly in the domains of cognition (Chapter 1), motivation, health behavior, and criminal activity (Chapter 2). Two initiatives cover the domains of jobs and careers and of organizational change; in both cases special attention should be given to developing samples and sites based not on households but on firms, agencies, voluntary affiliations, and occupational strata (Chapter 3). At the international level, new data collection initiatives are recommended on international economic transactions, migration, shifts in religious participation, and other processes parallel to ones under intensive study in the United States (Chapter 4).

The new data collections should be based principally on sample survey methods, but the value of these core collections can be greatly enriched by incorporation of research efforts using ethnographic, archival, and field experimental methods. Proposals for large-scale data collections require a two-track review process: one focusing on substantive scientific significance, in the same context as smaller-scale proposals; the other on an evaluation in terms of the distinct design criteria appropriate to large-scale studies. We recommend devoting an increment of $40 million annually to new, large-scale, multiuser data collections, involving an array of methods, in these areas.

Interdisciplinary Research Centers

New interdisciplinary research centers and facilities are not uncommon enterprises in the behavioral and social sciences. While some past starts have tended to segment along evolving disciplinary lines, other have maintained vital interdisciplinary programs for many years. We recommend a number of new initiatives for research centers, ranging from a major new international center to house demographic and related studies on developing countries, to a program of national research centers on motivational disorders and affective processes. We strongly encourage facilitation of major new center proposals, while sustaining existing centers of proven effectiveness. We recommend that $25 million annually be newly committed to the support of behavioral and social sciences research centers.

Investigator-Initiated Grants

In the recommendations for $170 million in new funds annually for research initiatives, we have considered the cultivation and support of research talent, the provision of technologically advanced instruments, and the building of new programs and organizations to generate data and centrally house and coordinate research activities. Still, the intellectual core and mainstay of behavioral and social sciences research—and of most scientific research—should be and will undoubtedly continue to be the small research group, including single investigators, who obtain support through discrete proposals evaluated by expert panels on their scientific merits. This reliance on investigator-initiated grants applies even to most of the research that is explicitly dependent on central data collections or technical facilities. We therefore recommend that the initiatives undertaken in each of the five research frontier areas include increments in investigator grant funds. The average size and duration of grants for small-group investigator research in most areas must be increased, even if the total number of grants awarded has to be constrained for a time in order to do so. Our estimates of the respective area requirements are detailed in Table 7-1 and add to $70 million.

Research Agency Changes

We recommend some changes in the ways that funding agencies evaluate, administer, and respond to their behavioral and social sciences research awardees. We recommend that the National Institutes of Health and the Alcohol, Drug Abuse, and Mental Health Administration extend and strengthen their portfolios of research on behavioral and social factors (in the etiology, prevention, and treatment of health problems) that are not specific to one disease. We recommend that the National Science Foundation and other funding agencies encourage the fusion rather than the separation of research on natural and artificial intelligence. We recommend staff increases in grant programs commensurate with the research initiatives proposed here, since the management and facilitation of interdisciplinary research demand more internal staff work, and such research is differentially threatened by staff reductions or individual grant-load increases. We recommend that the overall place and role of the behavioral and social sciences in the administrative arrangements of federal agencies must be critically reappraised with the intention of ensuring continuous high-level understanding of these fields' scientific needs and opportunities.

CONCLUSION

The total new funding recommended here is $240 million annually in current (1987) dollars, which we believe should be achieved within three or four years. In fiscal 1987, the total federal expenditure on behavioral and social sciences research—basic and applied, internal and external—came to about $780 million. About one-third of that was classified as basic research. Considering the small fraction of this increment that can come from the private sector or state government, we are recommending roughly a 30 percent overall near-term increase in federal support for behavioral and social sciences research. This increase would not be evenly distributed across all of the agencies that have a stake in such research. It is clearly weighted heavily toward what is usually classif d as basic research, and the most favorable opportunities lie more in some portfolios than in others. We estimate that the National Science Foundation should account for about $60 million of the new funding, and the National Institutes of Health and the Alcohol, Drug Abuse, and Mental Health Administration should receive about $80 million, with the rest divided among other departments and agencies and private sources.

We believe that the array of procedural innovations and the $240 million program of new investments outlined here and detailed in earlier chapters will ensure that the challenge of present research opportunities is vigorously met. Moreover, we believe these initiatives will best prepare the national research enterprise to explore new horizons of knowledge that cannot yet be seen, that lie beyond the veil that divides the present from the future.

Trends in Support for Research in the Behavioral and Social Sciences

FEDERAL SUPPORT

The major financial support for behavioral and social sciences research comes from the federal government and the operating budgets of doctorate-granting universities. In recent years, the financial circumstances of that research have largely been dominated by dramatic changes in levels of federal support. These changes in support for behavioral and social sciences are starkly different from those for other scientific disciplines, as shown in Figures A-1 and A-2.

Federal support for most scientific research has increased substantially in the past 15 years, with the value of regular, annual increases modulated by variations in the inflation rate. Considered in constant-dollar terms (that is, adjusting for inflation), the overall level of federal support for scientific research, exclusive of the behavioral and social sciences, was 36 percent higher in 1987 than in 1972.

Funds for the behavioral and social sciences, in contrast, have followed a roller-coaster-like track. In constant 1987 dollars, federal support declined from just over $1 billion in 1972 to $873 million in 1975, increased to a peak of about $1.1 billion in 1978 and 1979, fell sharply through 1982 to a low of

We are pleased to acknowledge the Division of Science Resources Studies of the National Science Foundation for providing unpublished data and other timely assistance in completion of this appendix.

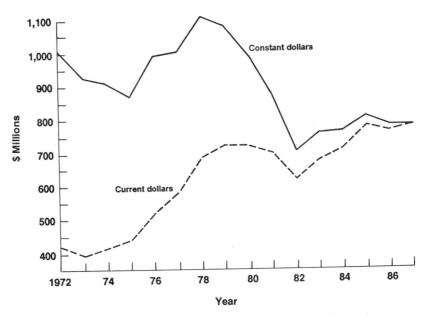

Figure A-1 Trends in federal support for behavioral and social sciences research (constant [1987] dollars). Source: Data from National Science Foundation, *Federal Funds for Research and Development.*

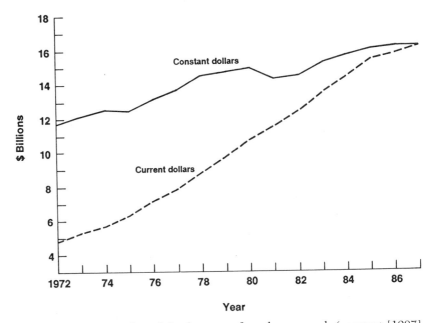

Figure A-2 Trends in federal support for other research (constant [1987] dollars). Source: Data from National Science Foundation, *Federal Funds for Research and Development.*

$705 million, then rebounded somewhat to the current level of $778 million. The 1987 level of support was 25 percent lower than the level in 1972.

The $778 million spent in 1987 includes external grants and contracts and internal research activities; about $270 million was classified as basic research and $508 million as applied research. The total is 4.6 percent of all federal expenditures for basic and applied research (about $16.8 billion). This percentage is a decline from a figure of 8 percent in 1972 and about 7 percent that was maintained (with some year-to-year fluctuations) from 1973 to 1979.

The Diversity of Federal Support

Although these highly aggregated numbers are indicative of overall trends, some diversity results from differences in missions and administrative practices among the many federal agencies that support extramural and perform intramural behavioral and social sciences research. This diversity affects the ways in which changes in funding levels are implemented and research is administered.

In federal research and development budgets, psychology is classified separately from social sciences, and the psychology classification also excludes neuroscience, which is classified with the life sciences. Basic research and applied research are classified separately, although the latter distinction is in many cases more a matter of the agency's mission than some intrinsic aspect of the research being supported. Figure A-3 shows the source of all federal funding for fiscal 1987, by agency.

In estimates of fiscal 1987 spending for basic research in psychology, roughly 60 percent (about $80 million) came from the U.S. Department of Health and Human Services, 26 percent ($36 million) from the U.S. Department of Defense, and 10 percent ($13 million) from the National Science Foundation. More than 90 percent of support for applied research came from Defense (47 percent, $87 million) and Health and Human Services (44 percent, $81 million); 5 percent ($9 million) came from the Veterans Administration.

In the social science categories, funding for basic research was more diverse: roughly 28 percent ($39 million) from Health and Human Services, 28 percent ($39 million) from the National Science Foundation, 19 percent ($26 million) from the Smithsonian Institution, 9 percent ($13 million) from the U.S. Department of Agriculture, 4 percent ($6 million) from the U.S. Department of Education, and 11 percent ($15 million) from other agencies. For applied social science research, 26 percent ($82 million) came from Health and Human Services, 20 percent ($70 million) from Agriculture, 20 percent ($64 million) from Education, and sizable though smaller amounts from the U.S. Department of Labor (7 percent, $7 million), independent agencies such as the National Aeronautics and Space Administration (4 percent, $11 million), and others (24 percent, $78 million).

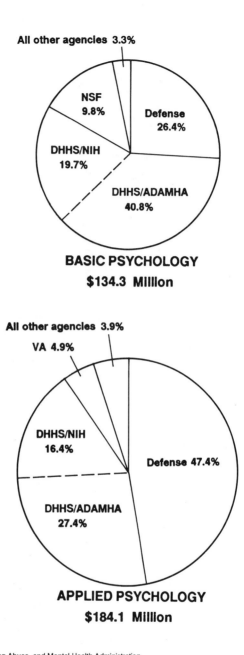

BASIC PSYCHOLOGY
$134.3 Million

All other agencies 3.3%
NSF 9.8%
Defense 26.4%
DHHS/NIH 19.7%
DHHS/ADAMHA 40.8%

All other agencies 3.9%
VA 4.9%
DHHS/NIH 16.4%
Defense 47.4%
DHHS/ADAMHA 27.4%

APPLIED PSYCHOLOGY
$184.1 Million

ADAMHA Alcohol, Drug Abuse, and Mental Health Administration
DHHS U.S. Department of Health and Human Services
EPA Environmental Protection Agency
HCFA Health Care Financing Administration
HUD U.S. Department of Housing and Urban Development
ITC International Trade Commission
NASA National Aeronautics and Space Administration
NIH National Institutes of Health
NSF National Science Foundation
SSA Social Security Administration
VA Veterans Administration

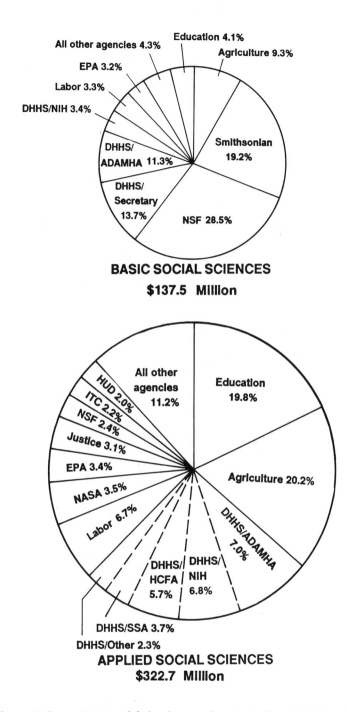

BASIC SOCIAL SCIENCES
$137.5 Million

Education 4.1%
Agriculture 9.3%
All other agencies 4.3%
EPA 3.2%
Labor 3.3%
DHHS/NIH 3.4%
DHHS/ADAMHA 11.3%
Smithsonian 19.2%
DHHS/Secretary 13.7%
NSF 28.5%

APPLIED SOCIAL SCIENCES
$322.7 Million

HUD 2.0%
ITC 2.2%
NSF 2.4%
Justice 3.1%
EPA 3.4%
NASA 3.5%
Labor 6.7%
All other agencies 11.2%
Education 19.8%
Agriculture 20.2%
DHHS/ADAMHA 7.0%
DHHS/HCFA 5.7%
DHHS/NIH 6.8%
DHHS/SSA 3.7%
DHHS/Other 2.3%

Figure A-3   Estimated federal expenditures in fiscal 1987 for basic and applied research in psychology and social sciences, by federal agency. Source: Data from National Science Foundation, *Federal Funds for Research and Development, Fiscal Years 1985, 1986, and 1987.*

In short, the Departments of Health and Human Services and Defense account for nine-tenths of federal research funding in psychology, while social science support is more diversified, although Health and Human Services provides one-fourth of all federal funds.

The National Science Foundation, with its specific mandate for nonmission-oriented research, has traditionally held a flagship role with respect to the social sciences, and the National Institute of Mental Health (now housed under the Alcohol, Drug Abuse, and Mental Health Administration) has shared this role with respect to psychology. But there is a strong presence in support of behavioral and social sciences research in other agencies, including national institutes devoted to aging, alcoholism and alcohol abuse, child health and human development, drug abuse, education, and justice; the Office of Naval Research, the Air Force Office of Scientific Research, and the Army Research Institute for the Behavioral and Social Sciences; the Cooperative State Research Service and Economic Research Service of the Department of Agriculture; the Social Security Administration, Health Care Financing Administration, and the Office of the Secretary in the Department of Health and Human Services; the Employment and Training Administration in the Department of Labor; the Veterans Administration; and the Smithsonian Institution. Also of great importance are federal agencies whose data collection activities are a foundation for work discussed in this report, especially the Bureau of the Census, the Bureau of Labor Statistics, the Bureau of Economic Analyses, the Bureau of Justice Statistics, the National Center for Health Statistics, and the National Center for Education Statistics. The diversity of departmental support and the preponderance of support for applied research testify to the payoffs gained from earlier investments.

National Science Foundation Support

The role of the National Science Foundation in federally funded research in universities and colleges is more pronounced than the overall figures cited above suggest, amounting to 25 percent of federal support for psychology and social sciences research on campuses and about 40 percent in the social sciences alone. Thus, changes in the nature and level of program support, while not strictly representative of government-wide action, have a major effect on academic research.

At the National Science Foundation, there are presently two divisions relevant to behavioral and social sciences research: Social and Economic Science, and Behavioral and Neural Sciences. (A division of Information Science and Technology was transferred in 1985 to the new Directorate for Computer and Information Science and Engineering and was renamed Information, Robotics, and Intelligent Systems; see discussion in Chapter 6.) As the following figures indicate, changes in the real-dollar budget for the Behavioral and Neural Sciences Division (BNS) have affected about equally the overall number of awards given and their average size. But in the Social and Economic Science Division (SES), much sharper changes in the budget have led to much more radical shifts in the size of grants awarded, with surprisingly small effects on the total number of grants. Because grant sizes have been reduced so drastically in SES, in several fields it has become very difficult to initiate new empirical studies. Instead, funds have been concentrated on sustaining relatively long-standing lines of theoretical work and extending a few longitudinal data sets.

Along with shifts to smaller grants—substantially smaller in the case of SES— there has been a change in the nature of the research personnel supported. Support for graduate students on research grants has decreased significantly, and faculty scientists receiving salary support through SES and BNS programs in 1985 were at salary levels roughly 30 percent higher, even after adjustment for inflation, than the faculty investigators on grants awarded just 3 years earlier. While some above-inflation rise in all faculty salaries has occurred, much of this increase represents a shift toward more research support for higher ranking faculty—or less to lower-ranking faculty—than previously.

The reduction in average grant sizes has occurred at the same time as major changes have taken place in the capabilities and costs of research equipment. Both the BNS and SES divisions participate in a multiuser equipment program, but the program has not been effective in meeting the overall equipment needs of researchers, partly due to specialized program requirements, partly due to the tendency of review panels to down-rate general-use equipment proposals in favor of individual-investigator proposals, and, as a reaction to these factors, partly due to limited numbers of proposals. The program also does not provide funds for the maintenance and operation of equipment or for hiring and training technical staff.

In addition to the reduction in the average grant size, the duration of awards has been shortened, presumably to enable administrators in some programs to maintain a relatively stable number of investigators in the face of declining budgets. But this change has a cost in research productivity because investigators more frequently have to write new proposals. This problem is by no means confined to behavioral and social sciences researchers, but it is especially acute for them. There are sound arguments for reversing this trend and moving to increase the duration of awards by 3 to 5 years, a direction in which some agencies, including the institutes of the Public Health Service, are now moving (see Chapter 6).

PRIVATE FOUNDATIONS

An important source of funding for the behavioral and social sciences is private foundations. The total of all private foundation support for behavioral

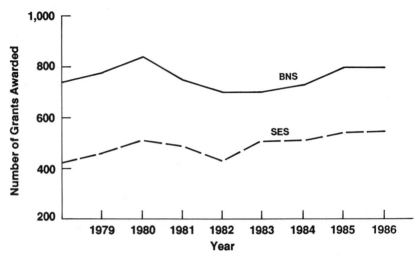

Figure A-4 Number of grants awarded for the Behavioral and Neural Sciences (BNS) and Social and Economic Science (SES) divisions of the National Science Foundation, 1978–1986. Source: Data from National Science Foundation, unpublished tabulations.

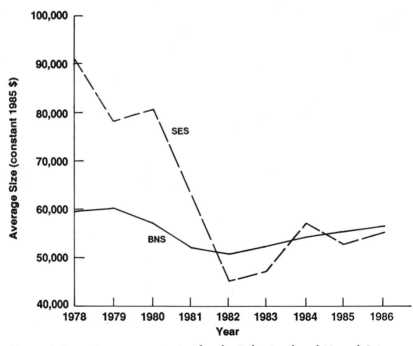

Figure A-5 Average grant size for the Behavioral and Neural Sciences (BNS) and Social and Economic Science (SES) divisions of the National Science Foundation, 1978–1986. Source: Data from National Science Foundation, unpublished tabulations.

258

Figure A-6   Total grant funding for the Behavioral and Neural Sciences (BNS) and Social and Economic Science (SES) divisions of the National Science Foundation, 1978–1986. Source: Data from National Science Foundation, unpublished tabulations.

and social sciences research is now about $60 million, or 8 percent of the amount received from the federal government. From the early 1920s until about 1960, private foundations were central in supporting behavioral and social sciences research; indeed, they were the principal providers of funds prior to the rapid growth of the federal presence beginning in the late 1950s. But private foundations were made subject to federal taxes during the 1960s, and substantial reductions occurred in the value of the awards they were able to give. Additionally, several very large foundations decided in the late 1950s and early 1960s to withdraw substantially from the support of basic research in favor of applied research and program development.

The role of the private sector thus changed from being an across-the-board presence in the behavioral and social sciences to providing limited support along selective program lines. Particularly notable are the Russell Sage Foundation (recently specializing in research on gender, legal processes, risk perception and management, and behavioral economics), the Sloan Foundation (cognitive science, economics), the Rockefeller Foundation (population, development), and the MacArthur Foundation (health and behavior). Although private foundations provide short-term selective program funding that injects significant resources for limited periods, they do not provide long-term commitment for the range of resource requirements needed to address many behavioral and social sciences research problems.

Working Group Members

Sensory and Perceptual Processes

NORMA V. GRAHAM (*Chair*), Department of Psychology, Columbia University
LINDA BARTOSHUK, Pierce Foundation, New Haven, Connecticut
ALBERT S. BREGMAN, Department of Psychology, McGill University, Canada
JULIAN HOCHBERG, Department of Psychology, Columbia University
AZRIEL ROSENFELD, Center for Automation Research, University of Maryland
MICHAEL STUDDERT-KENNEDY, Haskins Laboratory, New Haven, Connecticut

Psychobiology of Learning and Memory

RICHARD F. THOMPSON (*Chair*), Department of Psychology, Stanford University
CAROL BARNES, Department of Psychology, University of Colorado
THOMAS CAREW, Department of Psychology, Yale University
LEON COOPER, Department of Physics, Brown University
MICHELA GALLAGHER, Department of Psychology, University of North Carolina
MICHAEL POSNER, Department of Psychology, University of Oregon
ROBERT RESCORLA, Department of Psychology, University of Pennsylvania

---

Affiliations are those of members at the time the working groups began.

DANIEL SCHACTER, Department of Psychology, University of Toronto
LARRY SQUIRE, Veterans Administration Medical Center, and Department of Psychiatry, University of California, San Diego

Information and Cognitive Sciences

SAUL STERNBERG (*Chair*), University of Pennsylvania and AT&T Bell Laboratories, Murray Hill, New Jersey
FERGUS I. M. CRAIK, Department of Psychology, University of Toronto
JOHN JONIDES, Department of Psychology, University of Michigan
WALTER KINTSCH, Department of Psychology, University of Colorado
STEPHEN M. KOSSLYN, Department of Psychology, Harvard University
JAMES L. MCCLELLAND, Department of Psychology, Carnegie Mellon University
RAYMOND S. NICKERSEN, Bolt, Beranek, and Newman, Cambridge, Massachusetts

Language and Language Processing

FREDERICK J. NEWMEYER (*Chair*), Department of Linguistics, University of Washington
ANTONIO R. DAMASIO, Department of Neurology, University of Iowa Medical School
MERRILL GARRETT, Department of Psychology, Massachusetts Institute of Technology
MARK LIBERMAN, AT&T Bell Laboratories, Murray Hill, New Jersey
DAVID LIGHTFOOT, Linguistics Program, University of Maryland
HOWARD POIZNER, Laboratory for Language and Cognitive Skills, The Salk Institute for Biological Studies, San Diego, California
THOMAS ROEPER, Department of Linguistics, University of Massachusetts
ELEANOR SAFFRAN, Department of Neurology, Temple University Medical School
IVAN SAG, Department of Linguistics, Stanford University

Development of Cognitive and Social Competence

HERBERT L. PICK (*Chair*), Department of Psychology and Institute of Child Development, University of Minnesota
ANN L. BROWN, Department of Psychology, University of Illinois
CAROL DWECK, Laboratory of Human Development, Harvard University
ROBERT EMDE, Department of Psychology, University of Colorado
FRANK KEIL, Department of Psychology, Cornell University
DAVID KLAHR, Department of Psychology, Carnegie Mellon University

Ross D. Parke, Department of Psychology, University of Illinois
Steven Pinker, Department of Psychology, Massachusetts Institute of
Technology

Health and Behavior

David S. Krantz (*Chair*), Department of Medical Psychology, Uniformed
University of the Health Sciences
Leonard Epstein, Department of Psychiatry, University of Pittsburgh
School of Medicine
Norman Garmezy, Department of Psychology, University of Minnesota
Marsha Ory, Behavioral Sciences Research, National Institute on Aging,
National Institutes of Health
Leonard Pearlin, Department of Social and Behavioral Sciences, University
of California, San Francisco
Judith Rodin, Department of Psychology, Yale University
Marvin Stein, Department of Psychiatry, Mt. Sinai School of Medicine of
the City University of New York

Affect and Motivation

Hans C. Fibiger (*Chair*), Kinsmen Laboratory of Neurological Research,
University of British Columbia
William E. Bunney, Department of Psychiatry, University of California,
Irvine
Adrian Dunn, Department of Neuroscience, University of Florida
Martin L. Hoffman, Department of Psychology, University of Michigan
Carroll Izard, Department of Psychology, University of Delaware
Jerome Jaffe, Addiction Research Center, Baltimore City Hospital
Roy Wise, Department of Psychology, Concordia University
Steven Woods, Department of Psychology, University of Washington

Social Interaction

John F. Kihlstrom (*Chair*), Department of Psychology, University of
Wisconsin
Ellen Berscheid, Department of Psychology, University of Minnesota
John Darley, Department of Psychology, Princeton University
Reid Hastie, Department of Psychology, Northwestern University
Harold Kelley, Department of Psychology, University of California,
Los Angeles
Sheldon Stryker, Department of Sociology, University of Indiana

Gender Studies

NANCY M. HENLEY (*Chair*), Department of Psychology, University of California, Los Angeles

ROSE LAUB COSER, Department of Sociology, State University of New York, Stony Brook

JANE FLAX, Department of Political Science, Howard University

NAOMI QUINN, Department of Anthropology, Duke University

KATHRYN KISH SKLAR, Department of History, University of California, Los Angeles

Information and Decision Making

MARK J. MACHINA (*Chair*), Department of Economics, University of California, San Diego

ROBIN HOGARTH, Graduate School of Business, University of Chicago

KENNETH MACCRIMMON, Faculty of Commerce and Business Administration, University of British Columbia

JOHN ROBERTS, Graduate School of Business, Stanford University

ALVIN ROTH, Department of Economics, University of Pittsburgh

PAUL SLOVIC, Decision Research, Eugene, Oregon

RICHARD THALER, Graduate School of Business, Cornell University

Market Efficiency

OLIVER E. WILLIAMSON (*Chair*), School of Organization and Management, Yale University

JERRY HAUSMAN, Department of Economics, Massachusetts Institute of Technology

PAUL JOSKOW, Department of Economics, Massachusetts Institute of Technology

ROGER NOLL, Department of Economics, Stanford University

CHARLES PLOTT, Division of Humanities and Social Sciences, California Institute of Technology

VERNON SMITH, College of Business and Public Administration, Department of Economics, University of Arizona

DAVID WISE, John F. Kennedy School of Government, Harvard University

Jobs and Inequality

FRANK P. STAFFORD (*Chair*), Department of Economics, University of Michigan

JAMES BARON, Graduate School of Business, Stanford University

DANIEL HAMERMESH, Department of Economics, Michigan State University

CHRISTOPHER JENCKS, Department of Sociology, Northwestern University

MICHAEL REICH, Department of Economics, University of California, Berkeley

ROSS STOLZENBERG, Graduate Management Admissions Council, Santa Monica, California

DONALD J. TREIMAN, Department of Sociology, University of California, Los Angeles

Markets and Organizations

STANLEY REITER (*Chair*), Department of Economics, Northwestern University

KENNETH ARROW, Department of Economics and Hoover Institution, Stanford University

LANCE DAVIS, Division of Humanities and Social Sciences, California Institute of Technology

PAUL DIMAGGIO, Department of Sociology, Yale University

MARK GRANOVETTER, Department of Sociology, State University of New York, Stony Brook

JERRY GREEN, Department of Economics, Harvard University

THEODORE GROVES, Department of Economics, University of California, San Diego

MICHAEL HANNAN, Department of Sociology, Cornell University

ANDREW POSTLEWAITE, Department of Economics, University of Pennsylvania

ROY RADNER, AT&T Bell Laboratories, Murray Hill, New Jersey

KARL SHELL, Department of Economics, University of Pennsylvania

Collective Choice Institutions

WILLIAM H. RIKER (*Chair*), Department of Political Science, University of Rochester

JAMES S. COLEMAN, Department of Sociology, University of Chicago

BERNARD GROFMAN, School of the Social Sciences, University of California, Irvine

MICHAEL HECHTER, Department of Sociology, University of Arizona

JOHN LEDYARD, Department of Economics, Northwestern University

CHARLES PLOTT, Division of Humanities and Social Sciences, California Institute of Technology

KENNETH SHEPSLE, Department of Political Science, Washington University

Emergence of Social, Political, and Economic Institutions

DOUGLASS C. NORTH (*Chair*), Department of Economics, Washington University

ROBERT BATES, Division of Humanities and Social Sciences, California Institute of Technology

ROBERT BRENNER, Department of History, University of California, Los Angeles

JAMES COLEMAN, Department of Sociology, University of Chicago

ELIZABETH COLSON, Department of Anthropology, University of California, Berkeley

KENT FLANNERY, Department of Archeology, University of Michigan

VERNON SMITH, Department of Economics, University of Arizona

Urban Transformation and Migration

JOHN M. QUIGLEY (*Chair*), Graduate School of Public Policy and Department of Economics, University of California, Berkeley

ALEX ANAS, Department of Civil Engineering, Northwestern University

NILES HANSEN, Department of Economics, University of Texas

GEOFFREY HEWINGS, Department of Geography, University of Illinois

LARRY LONG, Center for Demographic Studies, Bureau of the Census, U.S. Department of Commerce

EDWARD S. MILLS, Department of Economics, Princeton University

RISA PALM, Department of Geography and Office of Academic Affairs, University of Colorado

Causes and Consequences of Demographic Change

SAMUEL H. PRESTON (*Chair*), Population Studies Center and Department of Sociology, University of Pennsylvania

ANSLEY J. COALE, Office of Population Research, Princeton University

KINGSLEY DAVIS, Hoover Institution, Stanford University

GEOFFREY MCNICOLL, Center for Policy Studies, Population Council, New York, New York

JANE MENKEN, Office of Population Research, Princeton University

T. PAUL SCHULTZ, Department of Economics, Yale University

DANIEL VINING, Population Studies Center, University of Pennsylvania

Family and Domestic Relations

JOHN MODELL (*Chair*), Department of History, Carnegie Mellon University

MARGARET CLARK, Department of Psychology, Carnegie Mellon University

WILLIAM J. GOODE, Department of Sociology, Stanford University

WILLIAM KESSEN, Department of Psychology, Yale University

RAYMOND SMITH, Department of Anthropology, University of Chicago

ROBERT WILLIS, Department of Economics, State University of New York, Stony Brook

Formal and Legal Processes

ROBERT KAGAN (*Chair*), Department of Political Science, University of
California, Berkeley
MARC GALANTER, School of Law, University of Wisconsin
JOEL HANDLER, School of Law, University of Wisconsin
SUSAN SILBEY, Department of Sociology, Wellesley College
STANTON WHEELER, School of Law, Yale University

Crime and Violence

ALFRED BLUMSTEIN (*Chair*), School of Urban and Public Affairs, Carnegie
Mellon University
RICHARD BERK, Social Process Research Institute, University of California,
Santa Barbara
PHILIP COOK, Institute of Public Policy and Department of Economics, Duke
University
DAVID FARRINGTON, Institute of Criminology, Cambridge University,
England
SAMUEL KRISLOV, Department of Political Science, University of Minnesota
ALBERT J. REISS, JR., Department of Sociology, Yale University
FRANK ZIMRING, School of Law, University of Chicago

Religion and Political Change

DANIEL LEVINE (*Chair*), Department of Political Science, University of
Michigan
LEONARD BINDER, Department of Political Science, University of Chicago
THOMAS BRUNEAU, Department of Political Science, McGill University,
Canada
JEAN COMAROFF, Department of Anthropology, University of Chicago
SUSAN HARDING, Department of Anthropology, University of Michigan
CHARLES KEYES, Department of Anthropology, University of Washington
ROBERT WUTHNOW, Department of Sociology, Princeton University

Culture and Ideology

JAMES W. FERNANDEZ (*Chair*), Department of Anthropology, Princeton
University
KEITH H. BASSO, Department of Anthropology, Yale University
KAREN BLU, Department of Anthropology, New York University
KENNETH BOULDING, Institute of Behavioral Sciences, University of Colorado
MURRAY EDELMAN, Department of Political Science, University of Wisconsin

STEPHEN GUDEMAN, Department of Anthropology, University of Minnesota
IVAN KARP, Department of Anthropology, Smithsonian Institution, Washington, D.C.
STEVEN KEARNEY, Department of Anthropology, University of California, Riverside
GEORGE MARCUS, Department of Anthropology, Rice University
DENNIS MCGILVARY, Department of Anthropology, University of Colorado
EMIKO OHNUKI-TIERNEY, Department of Anthropology, University of Wisconsin
WILLIAM SEWELL, Department of History, University of Arizona
ANN SWIDLER, Department of Sociology, Stanford University

Internationalization of Social, Economic, and Political Life

PETER B. EVANS (*Chair*), Department of Sociology, Brown University
BRUCE CUMMINGS, School of International Studies, University of Washington
ALBERT FISHLOW, Department of Economics, University of California, Berkeley
PETER GOUREVITCH, Department of Political Science, University of California, San Diego
JOHN MEYER, Department of Sociology, Stanford University
ALEJANDRO PORTES, Center for U.S.-Mexican Studies, University of California, San Diego
BARBARA STALLINGS, Department of Political Science, Yale University

International Crisis Management and Security Studies

ROBERT JERVIS (*Chair*), Department of Political Science, Columbia University
JOSHUA LEDERBERG, Office of the President, Rockefeller University
ROBERT NORTH, Department of Political Science, Stanford University
STEPHEN ROSEN, Center for International Affairs, Harvard University
JOHN STEINBRUNER, Foreign Policy Studies Program, The Brookings Institution, Washington, D.C.
DINA ZINNES, Department of Political Science, University of Illinois

Macroeconomic Policy

STANLEY FISCHER (*Chair*), Department of Economics, Massachusetts Institute of Technology
WILLIAM BEEMAN, Congressional Budget Office, Washington, D.C.
RUDIGER DORNBUSCH, Department of Economics, Massachusetts Institute of Technology
THOMAS SARGENT, Department of Economics, University of Minnesota

ROBERT SHILLER, Department of Economics, Yale University
LAWRENCE SUMMERS, Department of Economics, Harvard University

Social Studies of Modern Science and Technology

DANIEL J. KEVLES (*Chair*), Division of Humanities and Social Sciences, California Institute of Technology
JOHN HEILBRON, Office of History of Science and Technology, University of California, Berkeley
DOROTHY NELKIN, Program in Science, Technology, and Society, Cornell University
GLENN PORTER, Hagley Museum and Library, Wilmington, Delaware
MERRITT ROE SMITH, Program in Science, Technology, and Society, Massachusetts Institute of Technology
CHARLES ROSENBERG, Department of History and Sociology of Science, University of Pennsylvania

Social Knowledge Producing Institutions

THEDA R. SKOCPOL (*Chair*), Department of Sociology and Department of Political Science, University of Chicago
MARTIN BULMER, Department of Social Science and Administration, London School of Economics and Political Science, England
THOMAS JUSTER, Institute for Social Research, University of Michigan
DONALD MCCLOSKEY, Department of Economics and Department of History, University of Iowa
DOROTHY ROSS, Department of History, University of Virginia
ARNOLD THACKRAY, Department of History and Sociology of Science, University of Pennsylvania
CAROL WEISS, Graduate School of Education, Harvard University

Large-Scale Data Needs

WARREN E. MILLER (*Chair*), Department of Political Science, Arizona State University
JEROME CLUBB, Institute for Social Research, University of Michigan
MARTIN DAVID, Social Science Research Institute, University of Wisconsin
JAMES A. DAVIS, Department of Sociology, Harvard University
BRUCE RUSSETT, Department of Political Science, Yale University

Statistical Analysis

JOHN W. PRATT (*Chair*), Graduate School of Business, Harvard University
CLIFFORD CLOGG, Department of Sociology, Pennsylvania State University

BERT GREEN, Department of Psychology, Johns Hopkins University

MICHAEL HANNAN, Department of Sociology, Cornell University

JERRY HAUSMAN, Department of Economics, Massachusetts Institute of Technology

WILLIAM KRUSKAL, Department of Statistics, University of Chicago

DONALD RUBIN, Department of Statistics, Harvard University

RICHARD SAVAGE, Department of Statistics, Yale University

JOHN TUKEY, Department of Statistics, Princeton University

KENNETH WACHTER, Department of Statistics and Graduate Group in Demography, University of California, Berkeley

Measurement and Scaling

A. KIMBALL ROMNEY (*Chair*), School of the Social Sciences, University of California, Irvine

NORMAN BRADBURN, National Opinion Research Center, University of Chicago

J. DOUGLAS CARROLL, AT&T Bell Laboratories, Murray Hill, New Jersey

ROY G. D'ANDRADE, Department of Anthropology, University of California, San Diego

JOHN CLAUDE FALMAGNE, Department of Psychology, New York University

PAUL HOLLAND, Educational Testing Service, Princeton, New Jersey

LAWRENCE J. HUBERT, Department of Education, University of California, Santa Barbara

EDWARD E. LEAMER, Department of Economics, University of California, Los Angeles

Origins of Human Distinctiveness

GLYNN LL. ISAAC (*Chair*), Department of Anthropology, Harvard University

ROBERT BLUMENSCHINE, Department of Anthropology, Rutgers University

MARGARET CONKEY, Department of Anthropology, State University of New York, Binghamton

TERRY DEACON, Department of Anthropology, Harvard University

IRVEN DEVORE, Department of Anthropology, Harvard University

PETER ELLISON, Department of Anthropology, Harvard University

RICHARD FORD, Department of Anthropology, University of Michigan

KATHERINE MILTON, Department of Anthropology, University of California, Berkeley

DAVID PILBEAM, Department of Anthropology, Harvard University

RICHARD POTTS, Smithsonian Institution, Washington, D.C.

KATHY SCHICK, Department of Anthropology, University of California, Berkeley

MARGARET SCHOENINGER, Department of Anthropology, Harvard University
ANDREW SILLEN, Department of Anthropology, University of Pennsylvania
JOHN SPETH, Department of Anthropology, University of Michigan
NICHOLAS TOTH, Department of Anthropology, University of California,
    Berkeley
SHERWOOD WASHBURN, Department of Anthropology, University of
    California, Berkeley

Index

Index

275